POLYMER SCIENCE AND TECHNOLOGY

NEW DEVELOPMENTS IN POLYMERS RESEARCH

POLYMER SCIENCE AND TECHNOLOGY

Additional books in this series can be found on Nova's website under the Series tab.

Additional E-books in this series can be found on Nova's website under the E-book tab.

POLYMER SCIENCE AND TECHNOLOGY

NEW DEVELOPMENTS IN POLYMERS RESEARCH

EVAN O. BRADLEY
AND
MICHELLE I. LANE
EDITORS

Nova Science Publishers, Inc.
New York

Copyright © 2012 by Nova Science Publishers, Inc.

All rights reserved. No part of this book may be reproduced, stored in a retrieval system or transmitted in any form or by any means: electronic, electrostatic, magnetic, tape, mechanical photocopying, recording or otherwise without the written permission of the Publisher.

For permission to use material from this book please contact us:
Telephone 631-231-7269; Fax 631-231-8175
Web Site: http://www.novapublishers.com

NOTICE TO THE READER

The Publisher has taken reasonable care in the preparation of this book, but makes no expressed or implied warranty of any kind and assumes no responsibility for any errors or omissions. No liability is assumed for incidental or consequential damages in connection with or arising out of information contained in this book. The Publisher shall not be liable for any special, consequential, or exemplary damages resulting, in whole or in part, from the readers' use of, or reliance upon, this material. Any parts of this book based on government reports are so indicated and copyright is claimed for those parts to the extent applicable to compilations of such works.

Independent verification should be sought for any data, advice or recommendations contained in this book. In addition, no responsibility is assumed by the publisher for any injury and/or damage to persons or property arising from any methods, products, instructions, ideas or otherwise contained in this publication.

This publication is designed to provide accurate and authoritative information with regard to the subject matter covered herein. It is sold with the clear understanding that the Publisher is not engaged in rendering legal or any other professional services. If legal or any other expert assistance is required, the services of a competent person should be sought. FROM A DECLARATION OF PARTICIPANTS JOINTLY ADOPTED BY A COMMITTEE OF THE AMERICAN BAR ASSOCIATION AND A COMMITTEE OF PUBLISHERS.

Additional color graphics may be available in the e-book version of this book.

LIBRARY OF CONGRESS CATALOGING-IN-PUBLICATION DATA

New developments in polymers research / editors, Evan O. Bradley and Michelle I. Lane.
    p. cm.
Includes bibliographical references and index.
ISBN 978-1-61942-915-4 (soft cover)
1. Polymers--Research. I. Bradley, Evan O. II. Lane, Michelle I.
QD281.P6N443 2012
547'.7--dc23
                        2011053523

Published by Nova Science Publishers, Inc. † New York

CONTENTS

Preface		vii
Chapter 1	Raman Study of the Pressure and Temperature Induced Transformations in Crystalline Polymers of C_{60} *K. P. Meletov and G. A. Kourouklis*	1
Chapter 2	Bio-based (Co)Polyesters Containing 1,4-Cyclohexylene Units: Correlations between Stereochemistry and Phase Behavior *Annamaria Celli, Paola Marchese, Simone Sullalti and Corrado Berti*	63
Chapter 3	Thermo- and pH-Sensitivity of Poly (*N*-Vinylpyrrolidone) in Water Media *N. I. Pakuro, A. A. Arest-Yakubovich, B. I. Nakhmanovich and F. Kh. Chibirova*	103
Index		113

PREFACE

This new book presents and discusses current research in the study of polymer research. Topics discussed include a raman study of the pressure and temperature induced transformations in crystalline polymers of C60; eco-friendly (co)polyesters containing 1,4-cyclohexylene units and thermo-and pH-sensitivity of poly(N-vinylpyrrolidone) in water media.

Chapter 1 - High hydrostatic pressure causes a number of effects in fullerene polymers; the most interesting of them being further pressure-induced polymerization and subsequent structural phase transitions in partially polymerized fullerenes. The behavior of the phonon modes of the polymeric one-dimensional orthorhombic 1D-O phase, the two-dimensional tetragonal 2D-T and rhombohedral 2D-R phases of C_{60} have been studied as a function of pressure, up to ~30 GPa, at room temperature. The 1D-O polymeric phase, even at small pressures, undergoes pressure-enhanced photo-induced transformation to a new polymeric phase characterized by twinned polymeric chains. The photo-transformed 1D-O polymer and two-dimensional 2D-R and 2D-T polymeric phases undergo irreversible structural transformations at different pressures to new cross-linked three-dimensional polymeric structures. The phonon spectra of the high-pressure phases provide strong indication that the fullerene molecular cage is preserved in the recovered phases. The decomposition of the 2D-R polymer of C_{60} during high temperature treatment leads to the initial face centered cubic structure of the fullerene C_{60} monomer.

Chapter 2 - Today there is an increasing interest to develop novel eco-friendly polymers, i.e. materials produced from renewable resources, with low energy consumption, non-toxic to the environment, in some cases also biodegradable and with good mechanical performances. In the field of aliphatic polyesters, novel (co)polymers, containing 1,4-cyclohexylene units, are very promising materials, as obtainable from biomass, environmentally degradable and characterized by good mechanical properties. Moreover, in these polyesters the cyclic units have two possible configurations (cis and trans) which strongly influence the final phase behavior. Indeed, the trans isomer is more rigid and symmetric than the cis. Highly symmetrical units tend to improve the chain packing with a consequent increment in crystallinity and crystalline perfection. On the other hand, the cis isomer introduces kinks into the main chain, which hinder the formation of stable crystals. Thus, at high trans content the polyesters are characterized by a relative high degree of crystallinity, whereas at low trans content the polymers are amorphous. Therefore, accordingly to the final cis/trans ratio, the phase behavior of the homopolymers and copolymers significantly changes and the stereochemistry of the cycloaliphatic units result to be a key factor to tailor the final thermal properties of the material. In this paper the properties of some homopolymers and copolymers, containing the 1,4-cyclohexylene units with different cis/trans ratio, are discussed just in terms of the correlations between stereochemistry and phase behavior.

Chapter 3 - In recent years, a number of polymers that undergo phase separation in water solutions on temperature rising are studied. These polymers are characterized by lower critical solution temperatures (LCST). Poly(*N*-vinylpyrrolidone) (PVP) is not thermo- or pH - sensitive under usual conditions. However, since this polymer is widely used, especially in medicine, several studies are dedicated to the problem of making this polymer stimuli-responsive, too. In the review, the phase behavior of PVP in water solutions under various conditions is covered. The phase behavior of PVP-containing copolymers and hydrogels are described. The effect of the addition of salts, including transition metal ones, on the PVP phase separation temperature is considered, the attention being paid to the different influence of anions and cations on this value. It is known that PVP readily forms complexes with many organic and inorganic compounds. Examples of such complex formation effects on cloud points of the polymer solutions are given. The phase behavior of PVP is compared with that of poly(*N*-vinylcaprolactam), a PVP close analog, which is a well-known thermosensitive polymer.

In: New Developments in Polymers Research ISBN: 978-1-61942-915-4
Editors: E. O. Bradley and M. I. Lane © 2012 Nova Science Publishers, Inc.

Chapter 1

RAMAN STUDY OF THE PRESSURE AND TEMPERATURE INDUCED TRANSFORMATIONS IN CRYSTALLINE POLYMERS OF C_{60}

K. P. Meletov[1] and G. A. Kourouklis[2]

[1]Institute of Solid State Physics of the Russian Academy of Sciences
142432 Chernogolovka, Moscow region, Russia
[2]Physics Division, School of Technology
Aristotle University of Thessaloniki
GR-540 06 Thessaloniki, Greece

ABSTRACT

High hydrostatic pressure causes a number of effects in fullerene polymers; the most interesting of them being further pressure-induced polymerization and subsequent structural phase transitions in partially polymerized fullerenes. The behavior of the phonon modes of the polymeric one-dimensional orthorhombic 1D-O phase, the two-dimensional tetragonal 2D-T and rhombohedral 2D-R phases of C_{60} have been studied as a function of pressure, up to ~30 GPa, at room temperature. The 1D-O polymeric phase, even at small pressures, undergoes pressure-enhanced photo-induced transformation to a new polymeric phase characterized by twinned polymeric chains. The photo-transformed 1D-O polymer and two-dimensional 2D-R and 2D-T polymeric phases undergo irreversible structural transformations at different pressures to new cross-linked three-dimensional polymeric structures. The phonon spectra of the high-pressure phases provide strong indication that the fullerene molecular cage is preserved in the recovered phases. The decomposition of the 2D-R polymer of C_{60} during high temperature treatment leads to the initial face centered cubic structure of the fullerene C_{60} monomer.

1. INTRODUCTION

Polymerization involving C_{60} molecules is a complicated process since carbon atoms are found in both sp^2 and sp^3 hybridized states and therefore many possibilities appear for the formation of bonds due to existence of 30 unsaturated double C = C bonds in the fullerene molecular cage. It has been established that the polymerization process can be initiated by photons of visible or ultraviolet light [1], alkali metal doping [2], [3] and high-pressure/high-temperature (HPHT) treatment of the pristine C_{60} [4], [5]. The covalent polymeric bonds are usually formed by the so-called [2 + 2] cyclo-addition reaction via the formation of four-membered rings between adjacent fullerene molecules. The intermolecular bonding leads to appreciable decrease of the intermolecular distance and deformation of the fullerene molecular cage, these effects result in the lowering of symmetry of the highly symmetric parent C_{60} molecule [1]. Photopolymerization is responsible for the existence of some fraction of polymeric material in any C_{60} specimen exposed to daylight illumination. The photopolymerized C_{60} material contains mainly its dimeric state, which is insoluble in commonly used solvents but reverts to the initial monomeric state by heating at ~ 500 K for a short time [1]. Pristine C_{60} material absorbs light very efficiently resulting in polymerization but this happens up to the radiation penetration depth, which is about 10 μm in thickness for visible light. Bulk quantities of C_{60} polymers have been available after the development of the HPHT polymerization technology, which give the opportunity to produce various types of crystalline polymers of C_{60} [4], [5]. The HPHT polymers of C_{60} have attracted considerable attention because of their interesting structure and promising properties related to their hardness [6]. The crystalline structure and the dimensionality of HPHT polymers depends strongly on the pressure (P) and temperature (T) treatment conditions. Thus, the C_{60} molecules form linear polymeric chains having orthorhombic crystal structure (1D-O) and/or dimers and higher oligomers at lower P and T. At intermediate P and T two-dimensional polymeric layers are formed which have either a rhombohedral (2D-R) or a tetragonal (2D-T) crystal structure, while at higher P and T the face centered cubic structures, based on three-dimensionally (3D) cross-linked polymerization of the material, are formed [1], [4]-[7]. In addition, the treatment of the pristine C_{60} material under high non-uniform pressure and high temperature leads to the creation of several disordered polymeric phases, the so-called ultrahard fullerite phases [8], [9]. Detailed X-ray studies of these phases have revealed their 3D polymeric character [10], [11].

The polymerization of C_{60} is effected by the destruction of a number of double C = C intramolecular bonds and the creation of intermolecular covalent bonds associated with sp^3-like fourfold coordinated carbon atoms in the fullerene molecular cage. The linear polymeric chains with 4 sp^3-like coordinated carbon atoms in the fullerene molecular cage are obtained for temperatures in the range 500-600 K and pressure higher than 1 GPa [5]. Parallel straight chains form two ordered orthorhombic structures that belong to pseudo-tetragonal *Immm* space group (pressure above 2-3 GPa) and orthorhombic *Pmnn* space group (pressure below 2 GPa) [12]. The planar polymeric layers, with 8 and 12 sp^3-like coordinated carbon atoms in the fullerene molecular cage, obtained at temperatures in the range 700-900 K and pressure 1.5-9 GPa, show tetragonal or trigonal unit cells [4], [5]. These layers, stacked in a close-packed arrangement, form tetragonal and rhombohedral structures [4], [5]. The tetragonal structure, usually observed at pressures below 5 GPa, can be stabilized in two types of stacking depending on the temperature/pressure path conditions; the *Immm* orthorhombic structure (actually tetragonal because the *a* and *b* axes are almost equal) and the $P 4_2/mmc$ tetragonal structure [12]. The calculated lattice energies of these structures are very close; nevertheless, the energy of the $P 4_2/mmc$ stacking is lower than that of the *Immm* stacking a fact that results in the growth of samples containing mainly the $P 4_2/mmc$ structure with some inclusion of the *Immm* structure [13]. The rhombohedral structure, usually observed at pressures higher than 5 GPa, is formed by two types of layer stacking while both structures are described by the $R\overline{3}m$ space group [12].

The three-dimensional cross-linked polymeric structures can be obtained at pressures higher than 9 GPa and temperatures of about 600-700 K. The strong interest in these polymeric structures arises primarily from the expected extremely high hardness of these materials [14]. The rather disordered crystal structure of these polymers is face centered cubic; it becomes more and more disordered as the treatment temperature is raised [12]. Theoretical investigation [15] has predicted that the 3D-polymerized C_{60} might be formed by the application of uniaxial pressure perpendicular to the polymeric layers of the 2D-T phase of C_{60} belonging to pseudo-tetragonal *Immm* space group. According to the density-functional calculations, polymerization will take effect at a lattice constant of c = 10.7 Å, which is attainable at a pressure of ~20 GPa, and results in the formation of a stable metallic phase having 24 sp^3-like and 36 sp^2-like hybridized carbon atoms per each C_{60} molecule [15]. Another theoretical study [16] predicted that uniaxial compression perpendicular to the chains in the 1D or to the polymeric layers in the 2D polymeric phases of C_{60} leads to the 3D polymerization with 52, 56 or even 60 sp^3-like coordinated carbon atoms per C_{60} molecular cage. These transformations are expected to take place at pressures lower than 14 GPa and the new phases are semiconducting with large bulk and shear moduli.

Our interest, over several years, has been focused on the experimental study of the pressure-induced transformations in crystalline polymeric phases of C_{60} and the structural stability of the new high-pressure phases by means of *in-situ* Raman spectroscopy using the diamond anvil cell (DAC) technique. The Raman scattering and infrared absorption spectra of various crystalline polymeric phases, prepared under carefully controlled conditions of HPHT treatment, have very rich and well defined structure. Their intensity distribution and peak positions differ significantly for the pristine C_{60}, 1D-O, 2D-R and 2D-T polymeric phases as has been shown by the detailed study of their optical spectra combined with their structural analysis [17]. In addition, the perturbations in the structure of the C_{60} cages, caused by external disturbances like pressure, temperature, chemical bond formation etc., are manifested in the phonon spectrum [18], [19] giving important information on the environment of the molecular cage. Therefore, Raman spectroscopy can be used effectively for the identification of various polymeric phases of C_{60} as well as for the *in-situ* monitoring of high-pressure induced effects and phase transformations in the fullerene-based class of materials. In the present chapter, we are going to summarize the results of our perennial experimental study of the effect of pressure and temperature on the prototype linear and planar polymeric forms of C_{60}. Our aim is twofold. First, we are going to present our systematic work concerning the C_{60} materials, specifically the ones we may call template or prototype materials, like the linear and planar polymeric forms with the intention to document the conditions of their formation, namely, the chemical and thermodynamic parameters that control the way they are formed. With Raman spectroscopy, it is actually possible to probe the mechanism of the bond formation in the polymerization processes as well as the degree of polymerization. Second, we would like to point the implications of this knowledge for the future, especially for the understanding of the newly emerged class of all carbon materials as well as for their applications. We believe that starting with the polymerization process of C_{60} additional interesting materials for applications are going to be produced which will exploit both carbon bond properties as well as their various forms of dimensionality. This becomes more appealing after the new insight on carbon materials provided by the graphene structure properties. We think that there may be in the future much more carbon materials in addition to graphene, which will explore and utilize the carbon bond properties, the symmetry as well as the dimensionality of these materials and use the polymerization of C_{60} in tailoring new materials for specific applications.

2. RESEARCH METHOD AND EXPERIMENTAL TECHNIQUE

The two-dimensional polymeric samples were prepared from sublimed 99.99% pure C_{60} powder pressurized in a piston/cylinder type high-pressure device and kept at high pressure/high temperature for a certain time. The synthesis of the 2D-T polymer of C_{60} was performed at pressures in the range 2.3-2.5 GPa and at a temperature of about 820 K similar to the procedure described in [20]. X-ray analysis of the samples, after the high-pressure/high-temperature treatment, confirmed that the crystal structure of the polymer is tetragonal (a = b = 9.082 Å and c = 14.990 Å). Detailed analysis has shown that it is described better in a space group *Immm* rather than $P\,4_2/mmc$ due to a possible high degree of stacking disorder in the direction perpendicular to the polymeric layers [12]-[13]. The 2D-R polymer of C_{60} was obtained by subjecting pristine C_{60} powdered material to a pressure of P≈5 GPa at a temperature of 773 K. The X-ray analysis, after the high-pressure/high temperature treatment, confirmed that the crystal structure of the polymeric sample is rhombohedral (space group: $R\bar{3}m$, a = 9.22 Å and c = 24.6 Å) [4]. The samples of the 1D-O polymer were prepared by pressurizing of pure C_{60} powder at 1.2 GPa and 573 K in a "toroid"-type high-pressure device. The X-ray diffraction analysis has confirmed that the samples have the orthorhombic packing of linear polymeric chains (space group *Pmnn*: a = 9.098 Å, b = 9.831 Å, and c = 14.72 Å) [21].

Raman spectra were recorded using a triple monochromator (DILOR XY-500) equipped with a CCD liquid-nitrogen cooled detector system. The spectra were taken in the back-scattering geometry by the use of a micro-Raman system comprising an OLYMPUS microscope equipped with objectives of 100× and 20× magnification and a spatial resolution of ~1 μm and ~3.5 μm, respectively. The spectral width of the system was ~3 cm^{-1}. The 514.5, 488 and 647.1 nm lines of the Ar$^+$ and Cr$^+$ lasers were used for excitation. The laser power kept lower than 2 mW, measured directly before the high-pressure cell, in order to avoid the polymer decomposition caused by laser heating effects. The photoluminescence (PL) spectra were recorded using a single monochromator JOBIN YVON THR-1000 equipped with a CCD liquid-nitrogen cooled detector system. The spectral width of the system was ~4 cm^{-1}. The 488 nm line of an Ar$^+$ laser was used for excitation of the luminescence spectra.

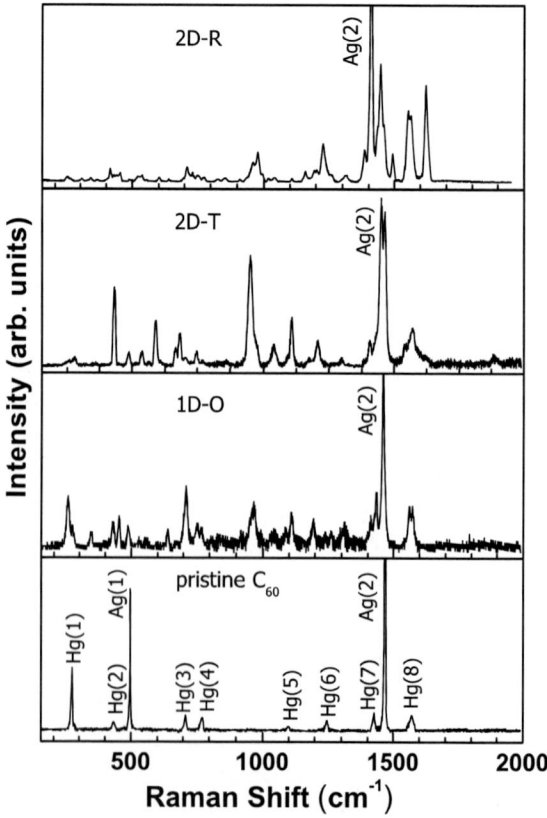

Figure 1. Raman spectra of various polymeric phases of C_{60} along with that of monomeric C_{60} recorded at normal conditions. The assignment given for the C_{60} refers to its molecular modes.

Measurements of the Raman and PL spectra at high pressures were carried out using the DAC of Mao-Bell type [22]. The 4:1 methanol-ethanol mixture was used as pressure transmitting medium and the ruby fluorescence technique was used for pressure calibration [23]. The Raman measurements at high temperature up to ~600 K were performed using a high temperature cell with a quartz window. The cell was equipped with a temperature controller unit that maintains temperatures up to 700 K with an accuracy of ±2 K. The heating and cooling rates in the temperature range of interest were less than ~15 K/min.

The polymeric samples with typical dimensions of ~100 μm were selected from the batch material by means of micro-Raman probing for their intense, clear and spatially uniform Raman response, characteristic for the 2D-T, 2D-R and 1D-O polymeric phases [17]. The Raman spectra of these polymers, in the frequency region 150-2000 cm^{-1} at normal conditions, are illustrated in Figure 1, along with that of the pristine C_{60}. The phonon frequencies, obtained by fitting the Raman data with Voigtian lines, are listed in Table 1. Polymerization results in the lowering of the molecular symmetry and therefore the Raman active fivefold-degenerated H_g modes are split and, in addition, new peaks may appear, originating from initially inactive modes of the C_{60} molecule. The assignment given in Table 1, is based on the icosahedral symmetry of the parent C_{60} molecule [17], [24], [25]. For simplicity, the irreducible representations of the C_{60} molecule are used to characterize the corresponding modes in the polymeric phases.

The most important probe in the Raman study of the polymeric phases of C_{60} is the behavior of the $A_g(2)$ pentagonal pinch (PP) mode of the C_{60} molecule. This mode is related to the in-phase stretching vibration of the five double C = C bonds, which involves tangential displacements with a contraction of the pentagonal rings and expansion of the hexagonal rings. Its frequency is very sensitive to any perturbation on the molecular cage and particularly to the breakdown of some bonds involved in the polymerization process. The PP mode is downshifted in all polymeric phases of C_{60}, as the breakdown of double C = C bonds and the subsequent formation of the intermolecular covalent bonds leads to lower intramolecular average bond stiffness. In the case of the *fcc* crystal structure of the C_{60} monomer, the $A_g(2)$ mode is observed at 1468 cm^{-1}. In the 1D-O polymer this peak is observed at 1458 cm^{-1} in excellent agreement with [26]. In the 2D-T polymer, the downshift of the PP mode is expected to be larger than in the 1D-O polymers as additional inter-molecular bonds are formed.

Table 1. Phonon frequencies of the various polymeric phases of C_{60} and their assignment based on the C_{60} molecular modes

Mode	1D(O) ω (cm^{-1})	2D(T) ω (cm^{-1})	2D(R) ω (cm^{-1})	Mode	1D(O) ω (cm^{-1})	2D(T) ω (cm^{-1})	2D(R) ω (cm^{-1})
$H_g(1)$	-	-	245	$F_{1g}(2)$	967	972	977
$H_g(1)$	255	257	267	$F_{2u}(4)$	-	-	1016
$H_g(1)$	273	279	308	$F_{2u}(4)$	1044	1039	1037
$H_u(1)$	345	-	342	$H_g(5)$	-	-	1042
$F_{2u}(1)$	-	-	366	$H_g(5)$	1087	1090	1078
$G_u(1)$	-	-	406	$H_g(5)$	1110	1108	1109
$H_g(2)$	429	432	415	$G_g(3)$	-	1176	1158
$H_g(2)$	-	-	438	$G_g(3)$	1193	-	1195
$H_g(2)$	453	-	451	$F_{2g}(3)$	-	1208	1204
$A_g(1)$	488	487	492	$H_g(6)$	1239	-	1224
$F_{1u}(1)$	528	-	520	$H_g(6)$	-	-	1230
$F_{2g}(1)$	-	537	532	$H_g(6)$	1260	-	1260
$F_{1g}(1)$	555	562	558	$G_g(4)$	1307	1299	1314
$F_{1g}(1)$	-	589	-	$H_g(7)$	1398	-	1385
$H_u(2)$	-	-	579	$H_g(7)$	1411	1405	-
$H_u(2)$	-	610	596	$H_g(7)$	1426	1428	-
$H_u(3)$	639	-	640	$H_g(7)$	1433	-	-
$H_g(3)$	-	666	-	$H_g(7)$	1448	-	-
$H_g(3)$	-	684	695	$A_g(2)$	1458	1446	1410
$F_{2u}(2)$	-	-	709	$F_{1g}(3)$	1464	1464	1461
$H_g(3)$	710	706	712	$F_{1g}(3)$	-	-	1495
$H_g(3)$	-	-	731	$F_{2g}(4)$	-	1543	-
$H_g(4)$	748	753	749	$H_g(8)$	1562	-	1554
$F_{2g}(2)$	-	-	767	$H_g(8)$	1575	1571	1563
$H_g(4)$	771	772	776	$H_g(8)$	-	-	1569
$F_{2u}(3)$	-	-	827	$G_g(6)$	-	1599	1621
$H_u(4)$	843	863	856	$G_g(6)$	-	1624	1627
$H_u(4)$	917	-	868	$2 \times G_g(2)$	-	1888	-
$G_g(2)$	955	950	958				

We observe indeed two peaks at 1446 and 1464 cm^{-1} in good agreement with [17]; the lower energy peak has been proposed to be the PP mode for the case of a 2D-T polymer, while the 1464 cm^{-1} peak has been attributed to the PP mode of C_{60} dimers [27]. However, from complementary analysis of Raman and infrared spectra of the 2D-T phase, it was concluded that the presence of an appreciable amount of dimers in the system is unlikely [6]. Therefore, this peak is of another origin; a possibility may be the activation of an initially inactive mode in I_h symmetry $\{F_{1g}(3)\}$, which becomes active in the D_{2h} symmetry of 1D-O and 2D-T phases. Finally, in the case of the 2D-R phase the $A_g(2)$ mode exhibits the strongest softening and is observed at 1408 cm^{-1}. Note, that a weak peak at 1446 cm^{-1} is related most likely to small inclusion of the 2D-T polymer in the samples used for Raman measurements. Thus, the synthesis of 2D-R polymer requires carefully controlled pressure/temperature conditions to minimize the content of tetragonal phase in the samples.

3. RAMAN SPECTRA OF THE 2D-T POLYMER AND PRESSURE INDUCED PHASE TRANSITIONS

The Raman spectra of the 2D-T polymer of C_{60} at various pressures up to 27.5 GPa and room temperature, in the frequency region 180-2050 cm^{-1}, are illustrated in Figure 2(a). The spectra were recorded upon pressure increase, while the spectral region around the strong triply degenerated T_{2g} mode of diamond, appearing at 1332 cm^{-1} at ambient pressure [28], is omitted. Apart from the pressure shift of the Raman bands, the first spectrum, recorded at 1.1 GPa, is similar to the typical Raman spectrum of the 2D-T polymeric phase at ambient conditions. The lowering of the cage symmetry, I_h of pristine C_{60} to D_{2h} in the 2D-T polymer, results in the splitting of the degenerated icosahedral intramolecular modes and to the activation of initially silent modes. Moreover, although the 2D-T phase retains the inversion center of the pristine C_{60} molecule, we cannot discard the possibility that imperfections in the crystal structure of the polymer and/or natural ^{13}C substitution may facilitate the appearance of some ungerade (u) modes in its Raman spectrum [29].

As the pressure increases, the Raman peaks shift to higher energies and their bandwidth gradually increases. The broadening of the Raman peaks is further enhanced for pressures higher than 10 GPa most probably due to the solidification of the pressure-transmitting medium. The pressure behavior of the Raman modes of the 2D-T polymer is fully reversible up to 12 GPa as it has been experimentally verified [29]. As can be clearly seen from Figure 2(a), the Raman peaks of the 2D-T polymer remain relatively narrow and well-resolved for pressures up to ~14 GPa, showing the stability of the used samples.

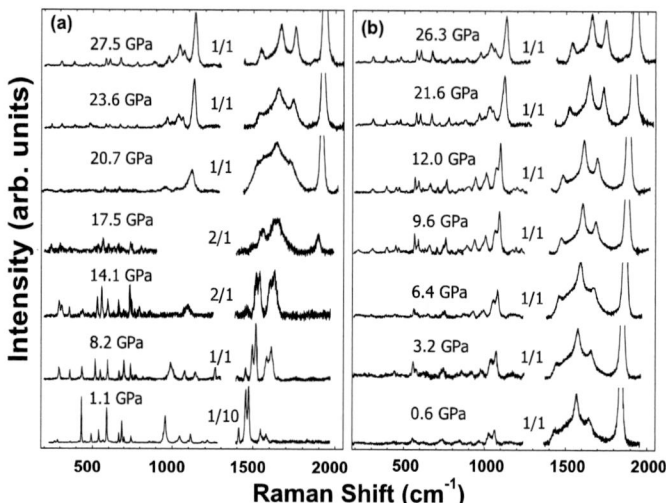

Figure 2. Raman spectra of the 2D-T polymer of C_{60} at 300 K and various pressures, recorded for (a) increasing and (b) decreasing pressure runs. The numbers 1/x indicate the relative scale of the spectra.

For pressures above 14 GPa the Raman peak bandwidths considerably increase, while the peak intensities decrease. In addition, the peak broadening is accompanied by a gradual enhancement of the background (not shown in Figure 2, as the Raman spectra are presented after background subtraction). The fluorescence from the 2D-T polymer of C_{60} is appearing in another energy region; therefore, this background is related, most likely, to the enhancement of strain and inhomogeneity within the sample induced at higher pressure.

Drastic changes in the Raman spectrum of the 2D-T polymer are first observed above 20 GPa; new distinct peaks appear in the spectrum while their intensities increase with increasing pressure. On the contrary, some of the initial Raman peaks disappear above this critical pressure. The Raman spectrum of the material for P ≥ 20 GPa differs significantly from that at lower pressure and the observed changes can be attributed to a new high-pressure phase. From Figure 2(a) it is clear, even for an applied pressure as high as ~27.5 GPa, that the Raman spectrum of the high-pressure phase is well resolved with relatively narrow peaks. Moreover, the frequency positions of the majority of the peaks in the new phase can be tracked back to the peaks observed in the initial 2D-T polymeric phase of C_{60}. This means that the C_{60} molecular cages are preserved at pressures higher than ~20 GPa as the Raman peaks in the high-pressure phase have their origin in the intramolecular cage vibrations. Figure 2(b) shows the Raman spectra of the material upon pressure release. The decrease of pressure, from ~27.5 GPa down to ambient conditions, results in the gradual shift of the Raman peaks of the high-pressure phase to lower energies. The release of pressure does not lead to any observable changes in the Raman intensity distribution and the high-pressure phase remains stable and is recovered at ambient conditions. The bottom spectrum in Figure 2(b) was recorded at ~0.6 GPa, while the sample was recovered in air after disassembling the DAC and tested again by means of micro-Raman probing.

The pressure dependence of the Raman modes of the 2D-T polymer in the initial (squares) and the high-pressure phases (circles) is shown in Figures 3 and 4. The open (solid) symbols denote data taken for increasing (decreasing) pressure runs. Solid lines are drawn to guide to the eye while arrows indicate the pressure increase or decrease. The mode assignment in these spectra refers to the irreducible representations of the parent C_{60} molecule following, in general, the same designation as in [17], and is given here only for the initial 2D-T phase of the polymer. The pressure dependence of the phonon frequencies of the 2D-T polymer shows linear behavior for almost all modes and is reversible for pressures up to 12 GPa [29]. However, the modes $H_g(1)$ and $A_g(1)$, demonstrate strong sub-linear pressure dependence.

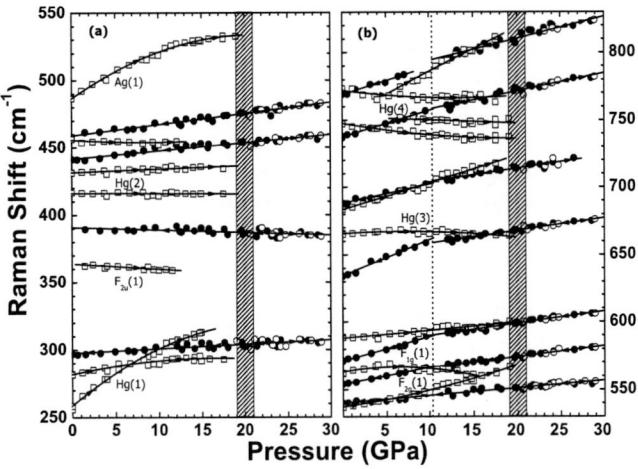

Figure 3. The pressure dependence of the Raman modes of the 2D-T polymer in the frequency regions 250-550 cm^{-1} (a) and 530-830 cm^{-1} (b). Squares and circles represent data taken for the 2D-T polymer in the initial and the high-pressure phase, respectively. The open (solid) symbols denote data taken for increasing (decreasing)

pressure runs. Shaded area around 20 GPa denotes the pressure range of the phase transformation.

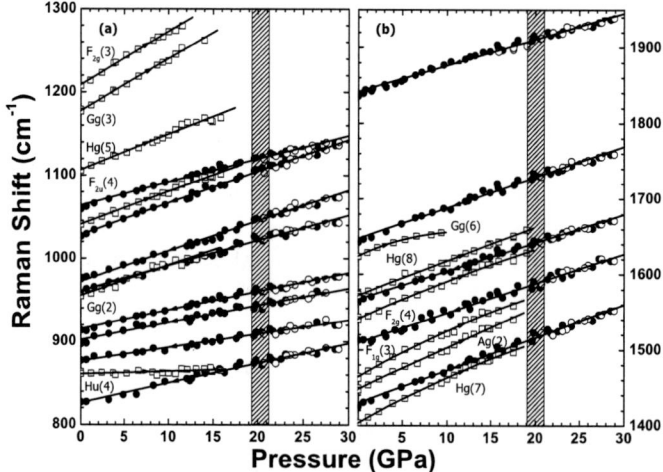

Figure 4. The pressure dependence of the Raman modes of the 2D-T polymer in the frequency regions 800-1300 cm^{-1} (a) and 1400-1950 cm^{-1} (b). Squares and circles represent data taken for the 2D-T polymer in the initial and the high-pressure phase, respectively. Open (solid) symbols denote data taken for increasing (decreasing) pressure runs. Shaded area around 20 GPa denotes the pressure range of the phase transformation.

The $A_g(1)$ mode is a "breathing" mode of the fullerene molecular cage and is associated with radial displacements in the atomic motions. The $H_g(1)$ mode is also related, to a large extent, to the radial displacements of the carbon atoms.

Thus, these two modes are characterized by out-of-plane displacements of carbon atoms and their sub-linear pressure dependence may be associated, in our opinion, with high anisotropy related to the van der Waals intermolecular bonding of adjacent 2D-polymeric layers and the covalent intermolecular bonding within the layers. Such a behavior is typical for the 2D polymeric layers and was also observed in the 2D-R polymer [30]. In addition, the $A_g(1)$ mode fully disappears for P ≥ 20 GPa which may be the result of the 3D polymeric bonding in the high-pressure phase, which quenches the "breathing" vibration of the fullerene molecular cage.

The frequencies of all the observed modes, in the high-pressure phase, increase with increasing pressure, except for the peak at 391 cm^{-1} which shows a small negative pressure slope. The pressure coefficients $\partial\omega_i/\partial P$ of the Raman modes in the initial phase of the 2D-T polymer vary from −1.2 up to 7.6 cm^{-1}/GPa, while the pressure slopes of the high-pressure phase are ranging from −0.2 up to 4.1 cm^{-1}/GPa. For comparison, the pressure coefficients of the Raman modes in the pristine C_{60} vary in a larger range from −4.1 up to 9.8 cm^{-1}/GPa. These data are in accordance with the fact that the polymerized materials become harder as the degree of polymerization increases [6], [15], [31]. It is interesting to note that the pressure coefficients of the Raman peaks at 1029 and 1064 cm^{-1}, associated with the sp^3-like coordinated carbon atoms, are comparable to that of the T_{2g} mode of the crystalline diamond (3.8, 2.8 and ~2.7 cm^{-1}/GPa, respectively) [32]. Finally, it is important to note that several Raman modes of the high-pressure phase, located in the frequency region 550-800 cm^{-1}, exhibit changes in their pressure slopes to higher values as the pressure decreases below 10 GPa. These changes may be related with the theoretically predicted [15] relaxation of the tetragonal lattice parameters in the high-pressure phase after pressure release. According to this study, the lattice parameter a of the high-pressure structure is expected to enlarge at ambient pressure by 0.3 Å becoming 9.4 Å. We think also that the relaxation of the lattice parameter in the recovered high-pressure phase is responsible for the softening of the 1040 cm^{-1} mode, in the initial 2D-T polymer, down to 1029 cm^{-1} in the new high-pressure phase (the low frequency split component).

The Raman modes of the high-pressure phase are related to those of the initial 2D-T polymer, as can be seen from Figures 3 and 4, showing that they originate from the C_{60} molecular cage vibrations. The nature of some phonon modes in the initial phase of the 2D-T polymer of C_{60}, in particular the Raman peak at ~1040 cm^{-1}, is related to the covalent intermolecular bonding within the 2D polymeric layers [6], [17]. More specifically, this is associated with the vibrations of the sp^3-like coordinated carbon atoms while the much lower frequency of this peak in comparison with that of the T_{2g} mode of diamond [28], is attributed to the different lengths of the sp^3-like bonds in the 2D-T polymer (1.64 Å) and diamond (1.54 Å). In the recovered high-pressure phase, this mode appears to have two components with frequencies 1029 and 1064 cm^{-1}. By assuming, that the high-pressure phase is related with the formation of a 3D polymeric phase of C_{60} proposed in [15], these two Raman peaks might be associated with the existence of two types of sp^3-like coordinated carbon atoms with slightly different bond lengths. Another important feature in the phonon spectrum of the high-pressure phase is the change in the region of the $A_g(2)$ pentagonal-pinch (PP) mode with respect to the pristine C_{60} and the initial 2D-T polymeric phase. The increase of the number of the sp^3-like coordinated carbon atoms in the high-pressure phase results in drastic changes in the PP-mode region. Namely, in the Raman spectrum of the high-pressure phase appear five strong peaks, with the most intense of them located at ~1842 cm^{-1}. Thus, the breakdown of a large number of double C = C bonds in the high-pressure phase leads to the quenching of the PP-mode, and as a result, a number of new Raman peaks appear, which are possibly related to the stretching vibrations of the remaining isolated double C = C bonds.

In Figure 5 the Raman spectrum of the 2D-T polymer, recorded at ambient conditions {Figure 5 (a)}, along with that of the high-pressure phase of the recovered material {Figure 5(b)} are illustrated for comparison. The spectra in Figure 5 were recorded out of the DAC and therefore it was possible to record also the spectrum of each material in the region of the T_{2g} mode of diamond. The recovered high-pressure phase of the 2D-T polymer exhibits a peculiar metastable behavior. Finally, in Figure 5(d) the Raman spectrum of another phase, which seems to be the lesser quantity - of the "detonated" sample, is given. This spectrum is rather weak, consisting of two relatively broad peaks at 1342 and 1591 cm^{-1}. There have been observed phases in C_{60}, treated at a pressure of 12.5 GPa and temperature higher than 700 °C [33], as well as in C_{60} treated at a pressure of 9.5 GPa and temperature higher than 1500 °C [9], having similar Raman spectra. The X-ray and microhardness studies of these phases have shown that they are related to disordered carbon phases having high density and hardness [9], [33], [34] associated with the break-down of C_{60} molecular cages and the formation of a cross-linked structure of graphite-like layers [9]. Indeed, the peak positions in the Raman spectra of these carbon phases, as well as the spectrum presented in Figure 5(d), are similar to those of the amorphous carbon containing a significant amount of sp^2 bonded carbon atoms [35] and to those of the microcrystalline graphite or the diamond-like carbon films, mostly comprising from sp^3 hybridized carbon atoms [36], [37]. The Raman peaks of this phase at 1591 cm^{-1} (C-band) is related to the stretching vibrations of the sp^2-like coordinated carbon atoms within the graphite layers, while the peak at 1342 cm^{-1} (D-band) appears in disordered carbon materials and is related to the vibrations of the sp^3-like coordinated carbon atoms.

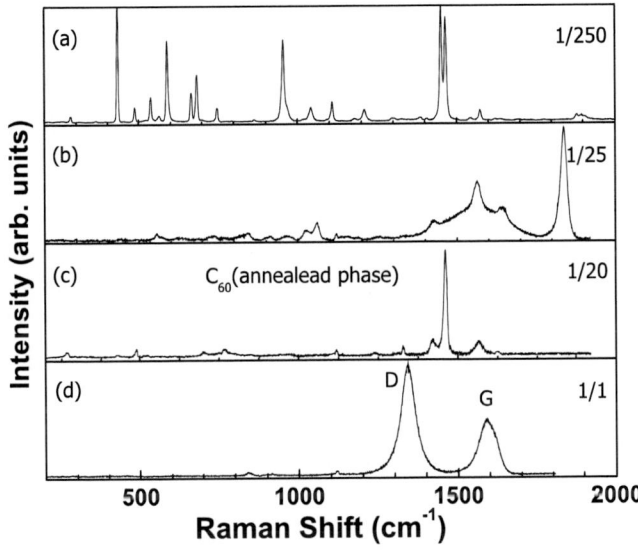

Figure 5. Raman spectra of the initial 2D-T polymer and of the recovered high-pressure phase after pressure release, recorded at ambient conditions. The numbers 1/x indicate the relative scale of the spectra. (a) The initial 2D-T polymer. (b) The high-pressure phase of the 2D-T polymer. (c) The main component among the pieces of the "detonated" sample identified as a mixture of the C_{60} monomer and dimer. (d) The "diamond-like" carbon phase identified among the pieces of the "detonated" sample.

The obtained experimental data provide a strong indication that the 2D-T polymer of C_{60} undergoes an irreversible phase transition above 20 GPa. The transformation takes place via a highly disordered pre-transitional state extending in a pressure range of ~4 GPa and having a rather diffuse Raman spectrum. The prominent Raman peaks of the high-pressure phase associated with the C_{60} molecular cage, as well as the irreversibility of the observed transformation, support the assumption of a further pressure-induced 3D-polymerization, which is rather a solid-state chemical reaction than a structural phase transition. The Raman spectrum of the high-pressure phase is dominated by a very strong Raman peak at ~1842 cm^{-1}, which cannot be related with any internal vibrational mode of the C_{60} molecular cage. The strong Raman peaks in the region of 1600 ÷ 1900 cm^{-1}, observed in some chemical compounds of carbon, are related to the stretching vibrations of isolated double C = C bonds [38]. In analogy to that, the strong peak at 1842 cm^{-1} can be attributed to the destruction of a number of double C=C bonds under further polymerization of 2D-T polymer leaving several isolated C=C bonds in the three-dimensional network of the C_{60} polymeric material. The results obtained by Raman measurements at high pressure have been verified by X-ray diffraction [39], [40], [41], where the crystal structure and stability at the high pressure of the 2D-T polymer were studied. The detailed study of the crystal lattice parameters of the 2D tetragonal and rhombohedral polymers of C_{60} at pressures up to 7 GPa by X-ray diffraction was performed in [40] but the crucial pressure range was not achieved. The first successful examination of the high-pressure Raman data performed in the X-ray study of the 2D-T polymer at pressure up to 37 GPa has revealed a transition at ~24 GPa associated with the formation of interlayer 3 + 3 cycloaddition along the body diagonal [39]. As a result, the transformation from 2D to 3D polymer was clearly demonstrated, while the structural model obtained in [39] differs from that predicted theoretically [15]. It is interesting to note, that the compressibility of the initial 2D-T polymer obtained by X-ray studies is highly anisotropic; the ambient pressure compressibilities along and perpendicular to the polymeric layers differ more that 20 times [39], [40]. After the transition to the high-pressure phase, the crystal structure remains tetragonal, the lattice parameter ratio c/a decreases abruptly from 1.66 to 1.36, while the anisotropy of compressibility disappears. Finally, the bulk modulus of the high-pressure 3D polymeric phase, determined from X-ray measurements, is 407 GPa which is slightly smaller that that of the diamond (443 GPa).

Thus, the X-ray study [39] has confirmed that the transformation found for the first time by Raman measurements [42], [43] is indeed of structural nature and is related to further 3D-polymerization of the initial 2D-T polymer. However, such a structural transition was not found in an earlier structural study of the 2D-T polymer at high pressure [41], which has shown a gradual amorphization of the initial polymeric phase at pressures higher than 20 GPa. A possible reason for this disagreement is the strong dependence of the pressure-induced 3D-polymerization of the 2D-T polymer on the structural details of the initial polymeric phase. As it was mentioned above, there are two crystal structures of the 2D-T polymer, characterized by the space groups of $P\,4_2/mmc$ and *Immm*. Since the initial 2D-T polymer in [39] was *Immm* with ~20% impurity of $P\,4_2/mmc$, the 3D polymerization, which is characteristic of *Immm* structure, did take place. However, in samples with $P\,4_2/mmc$ space group as a majority phase, a different transition is expected at a different pressure. Because this is a competing process with amorphization due to the nonhydrostaticity conditions in the diamond anvil cell, this may be the reason for the results observed by Leger *et al.* [41].

4. RAMAN SPECTRA OF THE 2D-R POLYMER AND PRESSURE INDUCED PHASE TRANSITIONS

The Raman spectra of the 2D-R polymeric phase of C_{60}, in the frequency region 100-2050 cm^{-1} and for various pressures up to ~30 GPa, are illustrated in Figure 6(a). The spectra were recorded upon pressure increase, while the spectral region around the strong triply degenerate T_{2g} mode of diamond, appearing at 1332 cm^{-1} at ambient pressure [28], is excluded. The background, which is gradually increasing with pressure, has been subtracted from the experimental spectra. The initial spectrum, taken at ~0.9 GPa, is similar to that recorded at ambient conditions and exhibits all typical Raman features of the 2D-R polymer [17]. The spectrum contains some additional peaks, indicated by arrows, related most likely with the presence of oligomers in the material. The Raman spectrum of the polymer is richer in structure than that of pristine C_{60} due to the splitting of the Raman active five-fold degenerate H_g phonon modes and/or to the activation of silent modes. Table 2 shows the correlation of the symmetry groups I_h (molecular symmetry of pristine C_{60}) and D_{3d} (molecular symmetry of C_{60} in the 2D-R polymer) [17]. As in the case of the 2D-T polymeric phase, defects and stresses in the crystal structure of the polymer, as well as natural isotopic ^{13}C substitution, may facilitate the appearance of some u-modes in the Raman spectrum of the material. The frequencies and the assignment of the Raman modes of the 2D-R polymer of C_{60} are shown in Table 3. The mode assignment refers to the irreducible representations of the C_{60} symmetry group [17]. As the pressure increases, the Raman peaks shift to higher energy and their bandwidth gradually increases.

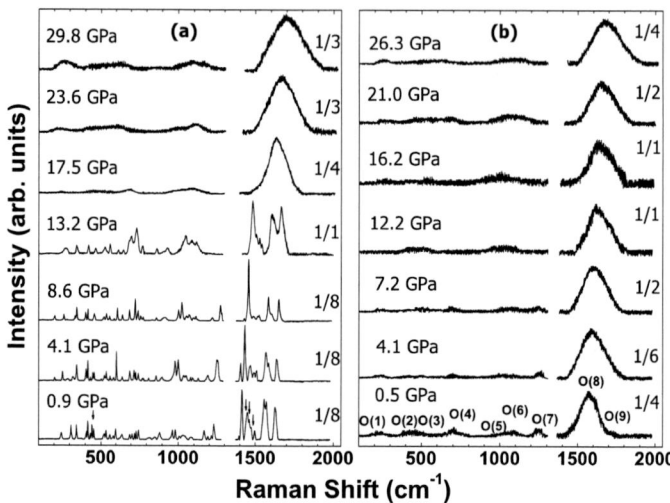

Figure 6. Raman spectra of the 2D-R polymer at room temperature and various pressures, recorded for (a) increasing and (b) decreasing pressure runs. The vertical arrows indicate peaks related to the presence of oligomers in the material. The numbers 1/x indicate the relative scale of the spectra.

Table 2. Correlations of the symmetry groups I_h (molecular symmetry in monomeric C_{60}) and D_{3d} (molecular symmetry in the 2D-R polymer of C_{60}) after [17]

I_h	D_{3d}	Splitting ($I_h \rightarrow D_{3d}$)
A_g	A_{1g}	$1 \rightarrow 1$
F_{1g}	$A_{2g} + E_g$	$0 \rightarrow 1$
F_{2g}	$A_{2g} + E_g$	$0 \rightarrow 1$
G_g	$A_{1g} + A_{2g} + E_g$	$0 \rightarrow 2$
H_g	$A_{1g} + 2E_g$	$1 \rightarrow 3$
A_u	A_{1u}	$0 \rightarrow 0$
F_{1u}	$A_{2u} + E_u$	$1 \rightarrow 2$
F_{2u}	$A_{2u} + E_u$	$0 \rightarrow 2$
G_u	$A_{1u} + A_{2u} + E_u$	$0 \rightarrow 2$
H_u	$A_{1u} + 2E_u$	$0 \rightarrow 2$

The pressure shift of the Raman peaks is accompanied by a gradual redistribution of the intensities among the various Raman modes: the relative intensities of the $H_g(3)$, $H_g(4)$, $G_g(2)$, $F_{1g}(2)$, $H_g(8)$, and $G_g(6)$ modes increase, while the intensities of the $H_g(1)$ and $A_g(1)$ modes decrease. These features in the pressure behavior of the Raman intensities are typical of the various polymeric phases of C_{60} and of some other fullerene-related materials [19], [29], [30]. The broadening of the Raman peaks is further enhanced above 10 GPa, most probably due to the solidification of the pressure-transmitting medium leading to pressure gradients.

Drastic changes in the Raman spectrum are first observed near 15 GPa, it becomes very diffuse and loses its fine structure in all frequency regions. This transformation in the Raman characteristics is preceded by a rapid decrease in the intensity of the $A_g(2)$ PP-mode which vanishes at pressures above 15 GPa. The Raman spectra at $P \geq 15$ GPa differ significantly from the spectrum of the 2D-R phase in the pre-transitional pressure range, exhibiting a considerable broadening of the Raman peaks, while the relative intensity distribution among the various phonon bands is preserved. The broad Raman features above $P \geq 15$ GPa, designated as $\Omega(1)$-$\Omega(9)$ in the fourth column of Table 3 can be tracked back to the initial 2D-R polymeric phase and seem to incorporate the corresponding group of the well resolved Raman peaks of this material at ambient pressure. The transformed material shows spatial uniformity as it was verified by its Raman response by probing various spots all over the surface of the sample.

Figure 6(b) shows the Raman spectra of the high-pressure phase of the 2D-R polymer upon pressure release. Decrease of pressure, from ~30 GPa down to ambient pressure, results in the gradual shift of the Raman peaks to lower energies without any observable changes in the Raman intensity distribution. The high-pressure phase remains stable down to ambient conditions. We should remark that the cycle of the pressure decrease lasted for a period of about one month. The number of the diffuse Raman bands and their positions in the recovered high-pressure phase are different from those in the initial 2D-R phase (see Table 3). The diffuse Raman spectra of the recovered material are typical of a highly disordered state, while the spectra taken at various sites of the sample are identical showing the spatial uniformity of the recovered material. The Raman spectrum of the high-pressure phase differs significantly from that of the amorphous carbon with respect to both the number of peaks and their frequencies. Our observations from several pressure runs show that the recovered sample is rather stable, not changing at ambient conditions at least for a period exceeding one week.

Table 3. Phonon frequencies and pressure coefficients of the 2D-R polymer of C_{60} and its "high-pressure" phase

2D-R polymer of C_{60}			"High-pressure" phase		
Mode[a]	ω_i cm^{-1}	$\partial\omega_i/\partial P$ cm^{-1}/GPa	Mode	ω_i cm^{-1}	$\partial\omega_i/\partial P$ cm^{-1}/GPa
$H_g(1)$	245	2.3	$\Omega(1)$	228	1.3
$H_g(1)$	267	2.8			
$H_g(1)$	308	3.4			
$H_u(1)$	342	0.6			
$G_u(1)$	406	-0.4	$\Omega(2)$	397	3.1
$H_g(2)$	415	0.2			
$H_g(2)$	438	2.2			
$H_g(2)$	451	0.9			
$A_g(1)$	492	1.1	$\Omega(3)$	470	4.7
$F_{1u}(1)$	520	0.1			
$F_{2g}(1)$	532	0.3			
$F_{1g}(1)$	558	-0.2			
$H_u(2)$	579	0.8			
$H_u(2)$	596	1.4			
$H_u(3)$	640	0.4			
$H_g(3)$	695	-0.5	$\Omega(4)$	705	-0.8
$H_g(3)$	709	-0.6			
$F_{2u}(2)$	712	1.8			
$H_g(3)$	731	-0.2			
$H_g(4)$	749	-0.2			
$F_{2g}(2)$	749	1.8			
$H_g(4)$	767	0.4			
$H_g(4)$	776	0.3			
$F_{2u}(3)$	827	1.0			
$H_u(4)$	856	0.8			
$G_g(2)$	958	5.0	$\Omega(5)$	1025	-1.4
$F_{1g}(2)$	977	5.4			
$H_g(5)$	1078	3.9	$\Omega(6)$	1086	0.3
$H_g(5)$	1109	4.2			
$G_g(3)$	1158	7.2			
$H_g(6)$	1224	5.8			
$H_g(6)$	1230	6.2	$\Omega(7)$	1244	0.6
$H_g(7)$	1385	5.4			
$A_g(2)$	1410	5.6			
$F_{1g}(3)$	1495	3.5			
$H_g(8)$	1554	3.7			
$H_g(8)$	1563	3.9	$\Omega(8)$	1568	4.3
$H_g(8)$	1569	4.3			
$G_g(6)$	1621	3.6			
$G_g(6)$	1627	3.8	$\Omega(9)$	1638	3.3

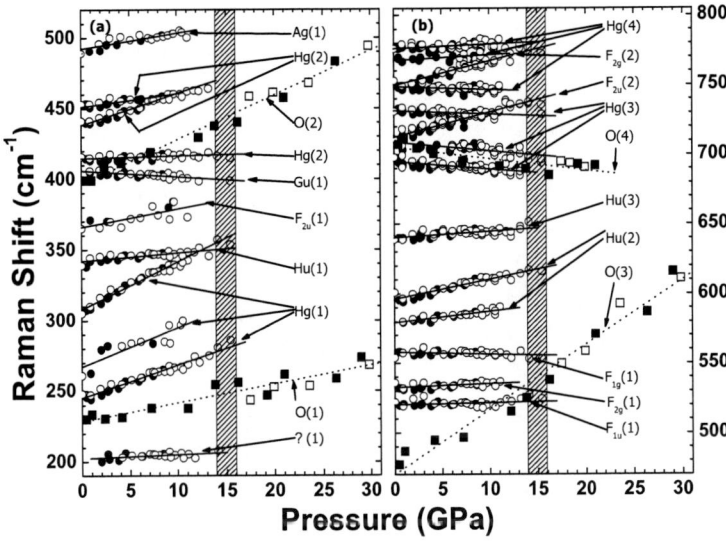

Figure 7. The pressure dependence of the Raman modes of the 2D-R polymer in the energy regions 190-520 cm^{-1} (a) and 470-810 cm^{-1} (b). Circles (squares) represent data taken for the 2D-R polymer in the initial phase (high-pressure phase). The open (solid) symbols denote data taken for increasing (decreasing) pressure runs. Shaded area around 15 GPa denotes the pressure range of the phase transformation.

The pressure dependence of the Raman modes of the 2D-R polymer of C_{60} for the initial (circles) and the high-pressure phase (squares) is shown in Figures 7(a) and 7(b) in the energy region 190–810 cm^{-1}. The open (solid) symbols denote data taken for upstroke (downstroke) pressure runs. Solid lines are guides to the eye. The mode assignment in these figures is related only to the initial 2D-R polymer and refers, as in Table 3, to the irreducible representations of the symmetry group of the parent C_{60} molecule (I_h symmetry) [44], following in general the same designation as in [17] with minor changes based on the present high-pressure Raman data. As it can be seen from Figures 7(a) and 7(b) (see also Table 3), the pressure-induced shift of the majority of the Raman modes of the initial 2D-R polymer is linear and positive, with exception of six modes with symmetries $G_u(1)$, $F_{1g}(1)$, $H_g(3)$, and $H_g(4)$, which display small negative pressure shifts. The pressure dependence of all Raman modes is reversible with pressure at least up to ~10 GPa, the highest pressure reached during the first experimental pressure run (solid circles in Figures 7(a) and 7(b)). The shaded area near 15 GPa indicates the pressure range where drastic changes in the Raman spectra take place. The transition to a new high-pressure phase results in the appearance of the diffuse

Raman features related to a highly disordered state. The frequencies of the bands in the high-pressure phase were defined with somewhat lower accuracy because of their diffuse nature. Their pressure dependence shows large dispersion and is positive for almost all modes, except for the $\Omega(4)$ mode, which shows a small negative pressure slope.

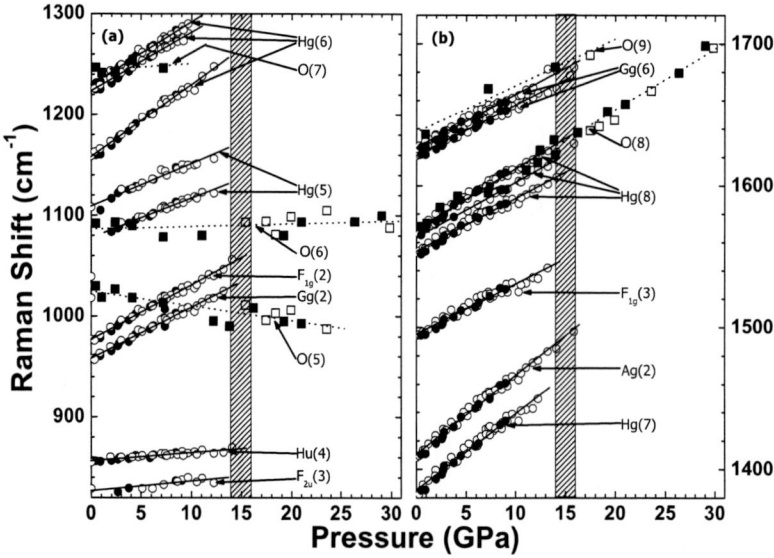

Figure 8. The pressure dependence of the Raman modes of the 2D-R polymer in the energy regions 820-1300 cm^{-1} (a) and 1380-1720 cm^{-1} (b). Circles (squares) represent data taken for the 2D-R polymer in the initial phase (high-pressure phase). The open (solid) symbols denote data taken for increasing (decreasing) pressure runs. Shaded area around 15 GPa denotes the pressure range of the phase transformation.

Figures 8(a) and 8(b) show the pressure dependence of the Raman modes of the 2D-R polymer for the initial (circles) and the high-pressure (squares) phases in the energy region 820-1720 cm^{-1}. All Raman modes of the initial 2D-R polymeric phase in this energy region show positive pressure shift with linear dependence, while this behavior is reversible with pressure for at least up to ~10 GPa. The shaded area near 15 GPa also indicates the pressure region where the irreversible transition to the disordered high-pressure phase takes place. The pressure dependence of the diffuse Raman bands in the high-pressure phase is also positive except for the $\Omega(5)$ mode. As it can be seen from Figures 7 and 8, the pressure dependence of the $\Omega(8)$ and $\Omega(9)$ modes, for the downstroke pressure run, is close to the pressure dependence of the corresponding group of modes $H_g(8)$ and $G_g(6)$ of the initial 2D-R polymer located in this frequency region. On the contrary, the pressure dependence of the $\Omega(1)$-$\Omega(7)$ modes differs significantly from that of the group of modes of the initial 2D-R polymer which are located in this frequency region. This may be partially related to the relatively high uncertainty in the determination of the frequency positions of the $\Omega(1)$-$\Omega(7)$ bands due to their low intensity in comparison with the more intense $\Omega(8)$-$\Omega(9)$ modes. The pressure coefficients of the Raman modes in the initial 2D-R phase range between –0.6 and 7.2 cm^{-1}/GPa, while those in the high-pressure phase vary between –1.4 and 4.7 cm^{-1}/GPa. At the same time, the pressure coefficients of the Raman modes in pristine C_{60} vary between –4.1 and 9.8 cm^{-1}/GPa. These results are compatible with the known experimental data and theoretical predictions that the polymerized fullerenes become harder as the degree of polymerization increases [6], [15], [31], [45].

Pressure-induced imperfections and structural defects may contribute to the broadening of the Raman bands at higher pressures. In order to explore such contributions the recovered samples were annealed under various temperature conditions. The Raman spectra of the annealed samples show that the material undergoes some transformation when subjected to a temperature higher than 550 K. Below this temperature no visible changes in the Raman response of the recovered sample were observed during the annealing procedure. The results of micro-Raman probing of the recovered sample at ambient conditions before and after annealing are shown in Figure 9. The Raman spectrum of the recovered sample after the pressure cycle, the Raman spectrum of the sample during partial annealing and after complete annealing at 550K are presented in Figures. 9(b), 9(c), and 9(d), respectively, along with the initial spectrum of the 2D-R phase in Figure 9(a). All spectra were recorded in the energy region 1350-1800 cm^{-1}, which includes the most intense Raman peaks. The Raman spectrum of the completely annealed material is spatially uniform and differs considerably from that of the high-pressure phase as well as from that of the initial 2D-R polymeric phase. This spectrum contains a relatively narrow and intense Raman band near 1464 cm^{-1}, related to the PP-mode in the case of a mixture of monomers and dimers of C_{60} and is similar to the spectra of various annealed polymeric phases of C_{60} [6], [43]. Note that the Raman response of the partially annealed sample in Figure 9(c) contains mainly the diffuse band related to the high-pressure phase, while the prominent Raman bands (shown by arrows) are related to the inclusion of the C_{60} dimers and monomers.

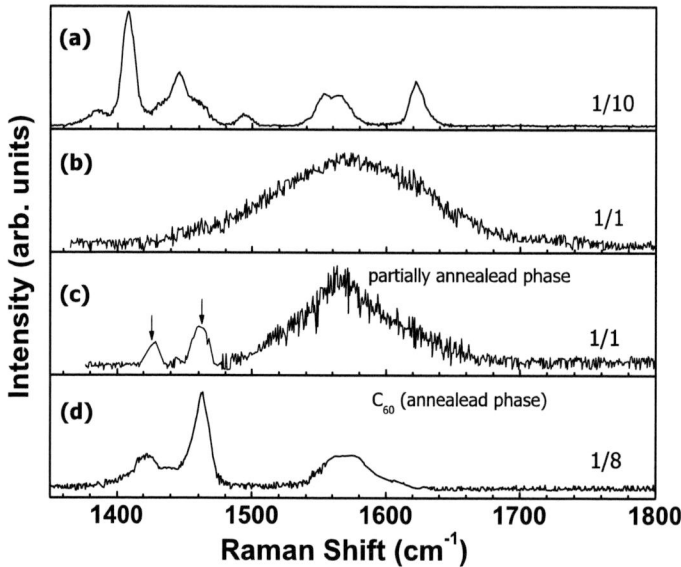

Figure 9. Raman spectra of the 2D-R polymer in the frequency region 1350-1800 cm^{-1} recorded at ambient conditions after release of pressure. (a) The initial 2D-R polymeric phase. (b) The "high-pressure" phase of the polymer. (c) The Raman spectrum of the "high-pressure" phase during partial annealing. Arrows indicate the Raman modes of the C_{60} dimer (d) The Raman spectrum of the "high-pressure" phase after being annealed at ~550 K. The spectrum is similar to that expected for a sample consisting of a mixture of C_{60} monomers and dimers.

The pressure behavior of the Raman modes provides strong indication that the 2D-R polymeric phase of C_{60} undergoes an irreversible transformation at a pressure of ~15 GPa. The initial well resolved Raman spectrum of the 2D-R polymer transforms to a diffuse one that is typical of a disordered state. The pressure dependence of the phonon modes is reversible up to ~10 GPa as it is unambiguously shown by the downstroke pressure data of the experimental run up to 10 GPa. At higher pressure, near 15 GPa, an irreversible transition to a new high-pressure phase takes place. The important feature in the pressure behavior of the phonon modes is the drastic changes in the region of the PP-mode related to the rapid decrease of the intensity of the $A_g(2)$ mode and the enhancement of the neighboring $H_g(8)$ and $G_g(6)$ modes in the pre-transitional pressure range. This behavior is reminiscent of an analogous behavior exhibited by these modes in the 2D-T polymeric phase of C_{60} before its further polymerization under high pressure [18], [43], [46]. The PP-mode in pristine C_{60} is related to the in-phase stretching vibration of the five double C = C bonds, which run away from the vertices of each pentagon in the fullerene molecular cage. The frequency of the PP-mode in the polymeric fullerenes decreases as the number of the polymeric covalent bonds per molecular cage increases (see Figure 1). As it was shown in the previous paragraph, the breakdown of a large number of double C = C bonds leads to the quenching of the initial PP-mode in the Raman spectrum of the high-pressure phase of the 2D-T polymer. The attenuation of the intensity of the PP-mode in the pre-transitional pressure range is related to the destruction of a number of double C = C bonds followed by subsequent creation of covalent links between molecules belonging to adjacent polymeric layers in the planar polymers of C_{60}. The application of high pressure decreases preferentially the distance between the polymeric layers due to the anisotropic compressibility of these materials, but the creation of covalent bonds is possible only between C atoms belonging in molecules of adjacent layers and having optimal relative orientations. The X-ray studies of the planar polymers of C_{60} at pressure up to 6 GPa have shown that the center-to-center distances of the nearest C_{60} cages between adjacent polymeric layers in the 2D-R polymer decreases more rapidly than those in the 2D-T polymer. However, the relative molecular orientations and atom-to-atom distances in the 2D-T polymer are more preferable for the formation of regular covalent bonds between molecules in adjacent polymeric layers [40].

Thus, we may assume that the new bonds in the highly compressed 2D-R polymeric phase of C_{60} are formed in a random way because of the presence of non-optimal molecular orientations. As a result, the new high-pressure phase exhibits a high degree of disorder characterized by random out-of-plane polymerization. This behavior differs from that of the 2D-T polymeric phase of C_{60} in which the *Immm* body-centered pseudo-tetragonal crystal structure provides optimal relative orientations of C_{60} molecules in adjacent sheets and leads to a high degree of regularity in the formation of out-of-plane covalent bonds [15]. This regularity in the covalent bonding is manifested in the Raman features of the high-pressure phase of the 2D-T polymer giving well-resolved and rich Raman spectra [42], [45]. It is important to note that inhomogeneity of the 2D-T samples that generally consist of a mixture of the *Immm* and *P* $4_2/mmc$ tetragonal phases may result also in spatially non-uniform Raman response of the high-pressure phase. The highly ordered high-pressure phase in compressed 2D-T polymer appears in a number of small islands having *Immm* structure and dispersed in the sample [43]. The rest of the sample having *P* $4_2/mmc$ structure is characterized by a rather diffuse Raman spectrum of the high-pressure phase, which is somewhat reminiscent that of the 2D-R polymer of C_{60}. Thus, in the primitive truly tetragonal structure with $P\ 4_2/mmc$ symmetry, the C_{60} molecules in adjacent polymeric layers do not have optimal relative orientations for the creation of covalent out-of-plane bonds in a regular way similar to that of the 2D-R polymer.

5. Photo- and Pressure-Induced Transformations in the Linear Orthorhombic Polymer of C_{60}

5.1. Photo-Induced Transformation

The Raman spectrum of the pristine 1D-O polymer recorded at ambient conditions is shown in Figure 10(a). This spectrum coincides with the earlier reported spectrum of the 1D-O polymeric phase, especially the number of the Raman active modes, the peak positions and their intensities are practically the same within the accuracy of measurements [17]. Figure 10(b) shows the Raman spectrum of the 1D-O polymer recorded in the first run of the high-pressure Raman measurements at a starting pressure of ~0.3 GPa obtained upon loading the DAC. The spectrum in Figure 10(b) has quite different structure: the majority of the bands are split and the total number of the peaks is increased. The same behavior, characteristic for pressure-induced transformation, is observed for a number of pressure runs with different specimens of the 1D-O polymer at any starting pressure obtained upon the sample loading into the DAC. The observed pressure-driven transformation is not typical: we have not been able to find the threshold pressure because it occurs even at pressure as low as 0.1 GPa, a different pressure generation technique, with higher pressure resolution in this range, is needed in order to check for this pressure threshold. Figures 10(c) and 10(d) show the Raman spectra of the planar 2D-T and 2D-R polymeric phases of C_{60}. The comparison of the Raman spectra of the transformed 1D-O polymer {Figure 10(b)}, to those of the 2D-T {Figure 10(c)} and 2D-R {Figure 10(d)} polymers clearly shows the differences between these polymeric forms in the number, position and relative intensities of the peaks.

The increase of the number of peaks testifies that the molecular symmetry of the C_{60} cage in the new phase of the 1D-O polymer is lower than in the pristine orthorhombic polymer (symmetry D_{2h}).

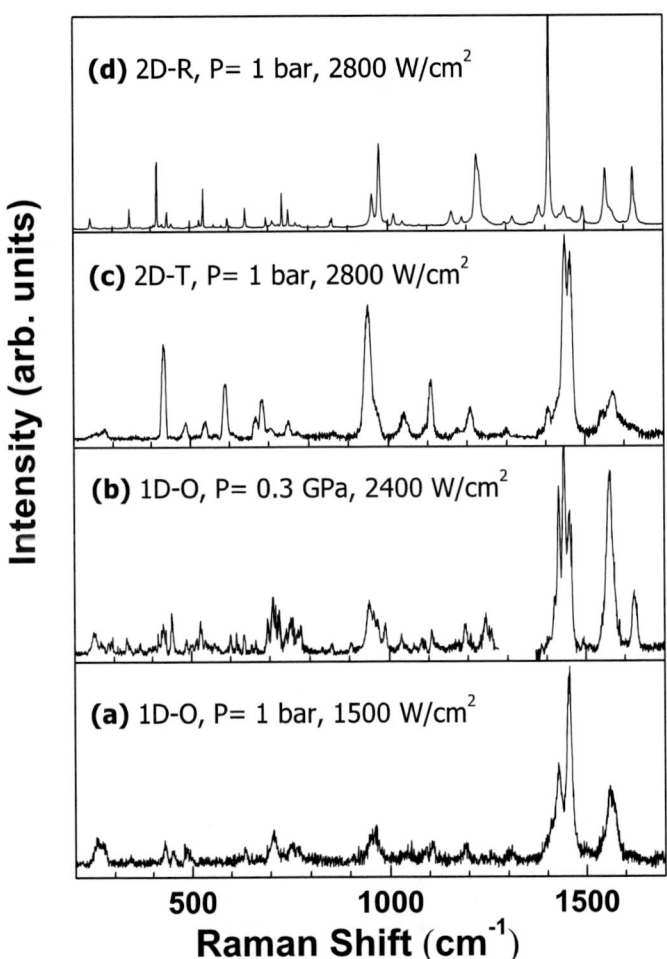

Figure 10. Raman spectra of the initial 1D-O polymer at normal conditions (a) and the transformed 1D-O polymer at 0.3 GPa (b) recorded with different excitation power densities. Raman spectra of the 2D-T (c) the 2D-R and (d) at ambient conditions are shown for comparison.

In the case of the cross-linking of the pristine 1D-O polymer, the lowering of symmetry could be associated with the formation of new inter-cage covalent bonds between the C_{60} clusters that belong to the neighboring polymeric chains in the structure of the orthorhombic phase. The distinction of the Raman spectrum of the transformed 1D-O polymer with respect to those of the 2D polymerized tetragonal and rhombohedral phases of C_{60} indicates that the new chemical bonds between neighboring polymeric chains of the 1D-O polymer are different from those of the cross-linked polymeric bonds in the planar polymers of C_{60}. The detailed data related to the Raman mode frequencies of these polymeric phases, as well as the Raman frequencies and the mode assignment of the pristine C_{60} are summarized in the Table 4.

Table 4. Phonon frequencies for the 2D-R, 2D-T, 1D-O and the transformed 1D-O polymeric phase of C_{60}. The corresponding values for monomer C_{60} are also included

2D-R polymer [30]		2D-T polymer [29]		1D-O transformed [47]		1D-O polymer [47]		Monomer C_{60} [48]	
Mode	ω_i (cm^{-1})	Mode	ω_i (cm^{-1})	Mode *	ω_i (cm^{-1})	Mode	ω_i (cm^{-1})	Mode	ω_i (cm^{-1})
$H_g(1)$	245	$H_g(1)$	259		248	$H_g(1)$	251	$H_g(1)$	273
$H_g(1)$	267		280		266		270		
$H_g(1)$	308				288				
$H_u(1)$	342				333		340		
$F_{2u}(1)$	366				366				
					389				
$G_u(1)$	406								
$H_g(2)$	415				411				
$H_g(2)$	438	$H_g(2)$	431		427	$H_g(2)$	425	$H_g(2)$	437
$H_g(2)$	451						450		
$A_g(1)$	492	$A_g(1)$	481		484	$A_g(1)$	486	$A_g(1)$	496
$F_{1u}(1)$	520				521	$\Omega(x)$	523		
$F_{2g}(1)$	532	$F_{2g}(1)$	536		527				
$F_{1g}(1)$	558	$F_{1g}(1)$	563		561				
$H_u(2)$	579	$F_{1g}(1)$	588						
$H_u(2)$	596				598				
		$\Omega(x)$	610		614				

2D-R polymer [30]		2D-T polymer [29]		1D-O transformed [47]		1D-O polymer [47]		Monomer C_{60} [48]	
Mode	ω_i (cm^{-1})	Mode	ω_i (cm^{-1})	Mode *	ω_i (cm^{-1})	Mode	ω_i (cm^{-1})	Mode	ω_i (cm^{-1})
$H_u(3)$	640				634	$H_g(3)$	635		
					654				
		$H_g(3)$	666		662				
$H_g(3)$	695				694				
$F_{2u}(2)$	709				707		707	$H_g(3)$	710
$H_g(3)$	712				722				
$H_g(3)$	731				739				
$H_g(4)$	749	$H_g(4)$	747		752	$H_g(4)$	752		
$F_{2g}(2)$	767								
$H_g(4)$	776		772		774		769	$H_g(4)$	774
$F_{2u}(3)$	827								
$H_u(4)$	856				853	$\Omega(x)$	843		
$H_u(4)$	868	$\Omega(x)$	864						
					903	$\Omega(x)$	897		
$G_g(2)$	958	$G_g(2)$	951		947	$\Omega(x)$	957		
					959				
$F_{1g}(2)$	977	$F_{1g}(2)$	970		969				
					987				
$F_{2u}(4)$	1016								
$F_{2u}(4)$	1037				1027	$\Omega(x)$	1034		
$H_g(5)$	1042	$\Omega(x)$	1041						
$H_g(5)$	1078	$H_g(5)$	1090		1082	$H_g(5)$	1082		
$H_g(5)$	1109		1107		1105		1105	$H_g(5)$	1100

Table 4. (Continued)

2D-R polymer [30]		2D-T polymer [29]		1D-O transformed [47]		1D-O polymer [47]		Monomer C_{60} [48]	
Mode	ω_i (cm^{-1})	Mode	ω_i (cm^{-1})	Mode *	ω_i (cm^{-1})	Mode	ω_i (cm^{-1})	Mode	ω_i (cm^{-1})
$G_g(3)$	1158	$G_g(3)$	1176						
$G_g(3)$	1195				1190		1190		
$F_{2g}(3)$	1204	$F_{2g}(3)$	1206		1205				
$H_g(6)$	1224								
$H_g(6)$	1230				1241	$H_g(6)$	1240	$H_g(6)$	1243
$H_g(6)$	1260				1257		1258		
$G_g(4)$	1314	$G_g(4)$	1299				1307		
$H_g(7)$	1385	$H_g(7)$	1404		1386	$H_g(7)$	1398		
$A_g(2)$	1410				1423		1416		
					1428		1429		1430	$H_g(7)$	1428
		$A_g(2)$	1447		1442		1442		
		$F_{1g}(3)$	1463		1455	$A_g(2)$	1457	$A_g(2)$	1470
$F_{1g}(3)$	1495								
$H_g(8)$	1554	$F_{2g}(4)$	1543		1559				
$H_g(8)$	1563	$H_g(8)$	1567		1559	$H_g(7)$	1560		
$H_g(8)$	1569	$G_g(6)$	1598				1575	$H_g(8)$	1575
$G_g(6)$	1621				1621		1621		
$G_g(6)$	1627								

*No assignment can be made.

To study the structural aspects of the observed transformation we have compared the X-ray diffraction pattern of the initial 1D-O polymer at normal conditions with that of the pressure-treated 1D-O polymer after pressure release. The pressure, up to 3 GPa, was generated in a "toroid"-type high-pressure cell while the duration of treatment was 10 minutes at 300 K. The X-ray diffraction pattern of the pristine material at ambient pressure and the pressure-treated 1D-O polymer after pressure release are shown in Figure 11(a) and Figure 11(b), respectively. As it can be seen from Figure 11, there are no significant differences in the X-ray diffraction patterns of the 1D-O polymer before and after pressure treatment. The positions of all observed peaks are the same, while small differences in peak intensities may be related to the powder material preparation. In addition, the *ex-situ* Raman spectrum of the pressure-treated 1D-O polymer taken after pressure release is the same with the spectrum of the initial polymer. Thus, the *ex-situ* X-ray and Raman data of the preliminary high-pressure treated 1D-O polymer do not show any changes in the crystal structure and the phonon spectrum of the material. On the contrary, the transformation is clearly observed in the *in-situ* high-pressure Raman study, which implies that the material transformation is related to the combined effect of laser irradiation and high pressure application. Therefore, the observed transition is related to pressure-enhanced photo-induced transformation of the material.

It is known that the HPHT 2D and 3D polymeric phases of fullerene are stable at ambient conditions and withstand the laser irradiation, which does not cause their further polymerization [6]. The main effect of laser irradiation at moderate and high intensities is the overheating of polymeric samples within the laser spot that may destruct the polymeric bonds and recover the initial monomer state.

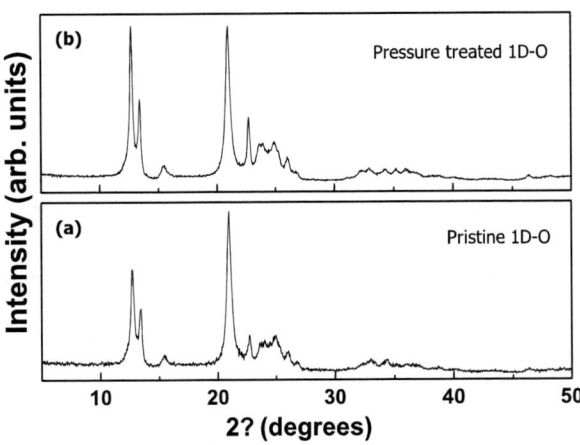

Figure 11. X-ray diffraction pattern at normal conditions of the initial orthorhombic phase of C_{60} (a), and the pressure-treated 1D-O polymer after pressure release (b).

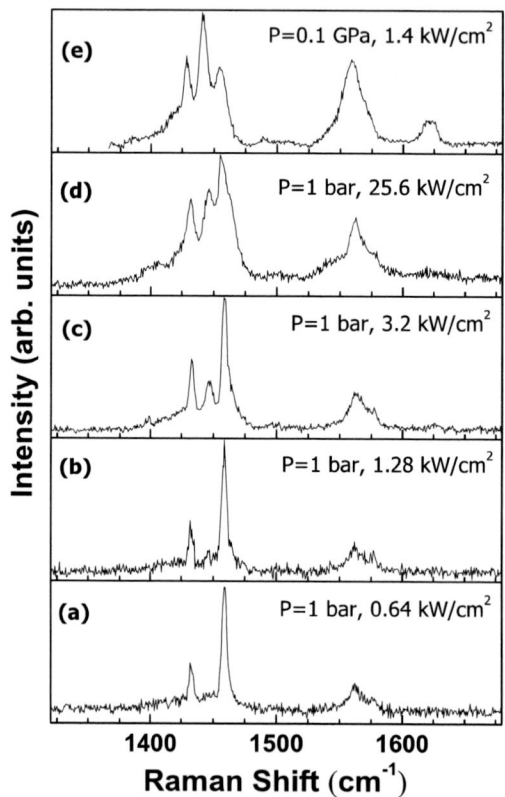

Figure 12. Raman spectra of the pristine 1D-O polymer recorded at ambient conditions and various excitation power densities of the 514.5 nm Ar+ laser line {panels (a)-(d)}. The spectrum of the transformed 1D-O polymer at 0.1 GPa is also included (e).

In order to check the stability of the linear orthorhombic polymer of C_{60} under laser irradiation at ambient conditions, we have measured the Raman spectra of the 1D-O polymer at various power densities using the 514.5 nm Ar+ laser line. The results of these measurements are presented in Figure 12. The spectrum in Figure 12(a), measured with a laser power density of 640 W/cm^2, is identical to the Raman spectrum of the pristine 1D-O polymer and does not show any traces of photo-induced transformation. The increase of the laser power density to ~1280 W/cm^2 (Figure 12(b)) does not affect the Raman features within the time scale of the experiment, while at a laser power density of ~3200 W/cm^2 a new Raman band appears near ~1446 cm^{-1}. Further increase of the laser power density to ~6400 W/cm^2 and subsequently to ~12800 W/cm^2 leads to the gradual enhancement of the new band intensity (not shown in the figure). At the same time, the two bands near 1433 cm^{-1} and 1563 cm^{-1} become stronger with respect to the main band at ~1458 cm^{-1}, which is attributed to the PP-mode of C_{60} in the pristine 1D-O polymer [17]. At the highest laser power density of ~25600 W/cm^2 {Figure 12(d)} the Raman bands broaden and shift to lower energies due to sample overheating, which results also in gradual degradation of the polymer under long time exposure. Note, that the latter spectrum is reminiscent of the Raman spectrum of the photo-transformed 1D-O polymer taken at P = 0.1 GPa and with a laser power density of 1400 W/cm^2 {Figure 12(e)}.

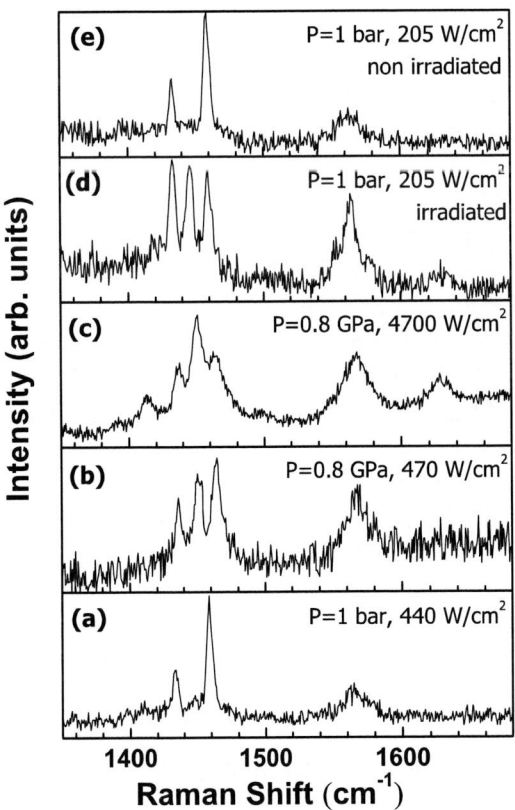

Figure 13. Raman spectra of the 1D-O polymer recorded at various pressures and excitation power densities of the 514.5 nm Ar$^+$ laser line. The pressure-assisted photo-polymerization starts at laser power density ~470 W/cm^2 (b) and intensifies at ~4700 W/cm^2 (c). After pressure release, the laser treated sites show Raman spectra typical of the photo-polymer (d), while untreated sites do not show photo-polymerization effects (e).

Figure 13 depicts the Raman spectra of the 1D-O polymer measured at various conditions of laser power density and pressure. The Raman spectrum of the 1D-O polymer shown in Figure 13(a) refers to ambient conditions at a laser power density of ~440 W/cm^2. Apparently, these experimental conditions do not cause changes related to photo-polymerization. At the pressure of 0.8 GPa the changes in the Raman spectrum appear at a laser power density of ~470 W/cm^2 {Figure 13(b)}. Further increase of the laser power density to ~4700 W/cm^2 leads to an almost instantaneous transformation of the material to a new phase {Figure 13(c)}. The Raman spectrum recorded in the DAC after pressure release shows the same features as those recorded under pressure of 0.8 GPa {Figure 13(d)}. This indicates that the new phase – appearing at the sample sites treated by laser irradiation at high pressure – remains stable at ambient conditions. On the contrary, the sample sites that were not irradiated by the laser beam at high pressure show typical Raman features of the pristine 1D-O polymer {Figure 13(e)}.

In order to avoid any influence of laser irradiation on the Raman spectra acquired after pressure release {Figure 13(d) and Figure 13(e)}, the measurements were performed at a laser power density of ~205 W/cm^2. This value is considerably lower than 3200 W/cm^2 at which the first Raman features related to photo-polymerization at ambient pressure appear. The obtained data imply that the irreversible changes in the Raman spectra of the pristine 1D-O polymer are related unambiguously to the laser irradiation of the samples, which leads to its further photo-induced polymerization. Note that the photo-induced polymerization of the 1D-O polymer, observed for the first time at ambient conditions for an HPHT type of C_{60} polymer, takes place at a laser power density of ~3200 W/cm^2, a value that exceeds more than two orders of magnitude 5 W/cm^2 reported for the photo-polymerization of the monomer C_{60} [1].

The application of high pressure enhances the process, resulting in the drastic increase of the photo-polymerization rate and in the subsequent reduction of the laser power density inducing the transformation. It is important to note that the molecules in the ground state cannot take part in the formation of C_{60} dimers via [2 + 2] cycloaddition reaction, which is the first step in the polymerization process. According to the Woodward-Hoffmann rule, the straightforward coupling of the C_{60} molecules in their ground state is not favorable due to the symmetry of the highest occupied orbital of C_{60} [49]. However, the molecular orbital of the excited state of C_{60}, being populated by light absorption, has favorable symmetry for dimer formation. On the contrary, the formation of dimers at high pressure takes place even at room temperature without light irradiation [50], which means that the highest occupied molecular orbital of C_{60} at high pressure is out of symmetry limitations on the pair interaction related to the Woodward-Hoffmann rule.

In view of this, the simultaneous effect of pressure and light irradiation can stimulate the polymerization process, which results in considerable increase of the polymerization rate. The results of this work confirm that the 1D-O polymer indeed becomes more sensitive to photochemical reactions when high pressure is applied. Concerning the possible mechanism of the high pressure photo-induced polymerization of the pristine 1D-O polymer, one can speculate that it may be similar to the mechanism of high pressure photo-induced transformation of other molecular carbon compounds with unsaturated bonds. In particular, the combined action of pressure and laser irradiation reduces the pressure threshold of the chemical transformation of crystalline benzene from 23 to 16 GPa [51]. According to this report, high pressure induces a distortion of the benzene ring that resembles the molecule in the first excited electronic state S_1. The change of molecular geometry relates to the pressure-induced mixing of the excited S_1 state with the S_0 ground state. That is, the distortion of the molecule at high pressure facilitates the photochemical transformation related to the selective pumping of the system in the S_1 excited state.

The structural features of the pressure-enhanced photo-induced polymerization of the 1D-O polymer may be attributed to the bonding among the linear polymeric chains. The lowering of symmetry, indicated by the appearance of new Raman peaks, can be associated with the formation of new covalent bonds between C_{60} molecules that belong to the neighboring polymeric chains in the structure of the orthorhombic phase. The distinction of the Raman spectrum of the transformed 1D-O polymer from those of the 2D polymerized tetragonal and rhombohedral phases of C_{60} is indicative of new chemical bonds in the transformed orthorhombic phase that are not the typical [2 + 2] cycloaddition bonds but rather single bonds between polymeric chains. The *in-situ* high pressure X-ray powder diffraction study has clearly shown that the 1D-O polymer transforms to a new polymeric phase, characterized by conjunction of adjacent linear polymeric chains [52]. This study has not revealed the detailed structure of the new polymeric phase since the low-resolution diffraction profiles along with the small number of peaks could not permit further refinement; nevertheless, additional synchrotron radiation X-ray diffraction experiments may clarify this issue.

5.2. Pressure-Induced Transformations

The stability at high pressure and additional pressure-induced transformations of the photo-transformed 1D-O polymer were investigated by means of Raman measurements at pressures up to 29 GPa. The Raman spectra of the photo-transformed 1D-O polymer of C_{60} at various pressures and room temperature are shown in Figure 14. Data recorded upon pressure increase are illustrated in Figure 14(a) whereas those recorded during pressure release in Figure 14(b). The spectral region around the strong triply degenerate T_{2g} mode of diamond, appearing at 1332 cm^{-1} at ambient pressure [28], is again excluded. The background, which increases slightly with pressure, has been subtracted from the spectra for clarity. The initial Raman spectrum of the photo-transformed 1D-O polymer, which consists of a large number of narrow and well-resolved peaks, demonstrates strong pressure dependence. As the pressure increases, the Raman peaks shift to higher energies and their bandwidth gradually increases. The broadening of the Raman bands is further enhanced above 10 GPa due to the solidification of the pressure-transmitting medium. The most important pressure effects are related to the changes in the number and intensities of the Raman active modes, as well as to their pressure coefficients. Namely, the relative intensities of the Raman modes in the 700 - 1100 cm^{-1} region gradually increase with pressure, while the most important changes were observed near ~15 GPa, where the Raman spectrum loses its fine structure in all frequency regions and becomes very diffuse. This transformation was preceded by a rapid decrease in the intensity of the peaks related to the $A_g(2)$ mode of the pristine C_{60} accompanied by a relative increase in the intensities of the $H_g(8)$ and $G_g(6)$ modes. The broad Raman features in the spectrum of the high-pressure phase above ~15 GPa can be tracked back to the photo-transformed 1D-O polymer of C_{60} and seems that is incorporating the corresponding group of the broad Raman bands of this phase. Despite the similarities in the diffused Raman bands, the spectrum of the high-pressure phase differs significantly from that of the amorphous carbon with respect to the number of peaks as well as to their position. Comparing the Raman spectrum of the high-pressure phase to that of the high-pressure phase of the 2D-T polymeric phase of C_{60} [42], [43], we observe that the former does not contain peaks in the high energy region, like the 1840 cm^{-1} peak observed in the latter phase. In addition, the new phase shows spatially uniform Raman response over all the surface of the sample, as it was documented by probing various places in the sample, a behavior which differs drastically from the 2D-T polymeric phase of C_{60} [42]. Note, that the broad

Raman features of the high-pressure phase in the photo-transformed 1D-O polymer of C_{60} resemble the Raman features of the disordered high-pressure phase in the 2D-R polymer, which was also observed above ~15 GPa [30]. The broad Raman bands of the disordered phase shift to lower energies, upon pressure decrease, without any observable changes in their intensity distribution {Figure 14(b)}. The high-pressure phase is recovered and remains stable for several hours at normal conditions.

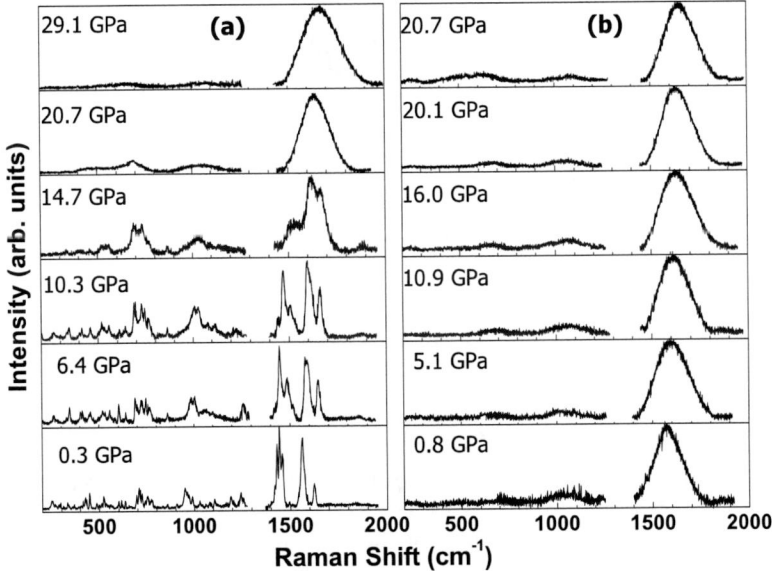

Figure 14. Raman spectra of the photo-transformed 1D-O polymer of C_{60} at room temperature and various pressures, recorded upon pressure increase (a), and upon pressure decrease (b).

The pressure dependence of the Raman mode frequencies of the photo-transformed 1D-O polymer of C_{60} in the energy regions 230 – 790 cm^{-1} and 840- 1720 cm^{-1} is shown in Figure 15(a) and Figure 15(b). Circles (squares) represent data recorded in a pressure run up to 12 GPa (29 GPa). Stars denote the Raman frequencies of the pristine 1D-O polymer at normal conditions. Moreover, open (solid) symbols represent data recorded upon pressure increase (decrease). The phonon frequencies obtained upon pressure increase coincide in the two different pressure runs within the accuracy of measurements. The pressure-induced shift of the majority of Raman modes is linear and positive, with exception of a few modes, which display small negative pressure shifts. The pressure coefficients vary between -0.4 cm^{-1}/GPa and 7.0 cm^{-1}/GPa. The pressure dependence of all Raman modes is reversible with pressure at least up to ~12 GPa, the highest pressure reached during the first pressure run.

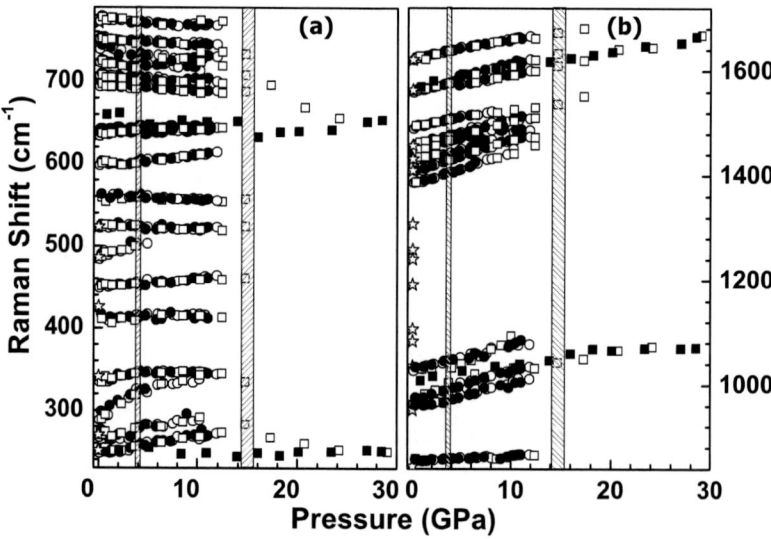

Figure 15. Pressure dependence of the Raman frequencies of the photo-transformed 1D-O polymer of C_{60} in the region 240-790 cm^{-1} (a) and 860 – 1720 cm^{-1} (b). Circles and squares represent data taken upon two different pressure runs up to 12 GPa and 29 GPa, respectively. Stars denote the Raman frequencies of the initial 1D-O polymer at normal conditions. The open (solid) symbols represent data recorded upon pressure increase (decrease). Dashed areas around 4 GPa and 15 GPa mark pressure values where the changes in the pressure-dependence were observed.

The dashed areas around 4 GPa and 15 GPa indicate the pressure ranges where the transformations of the Raman spectra are taking place. The pressure coefficients of some Raman bands change abruptly at ~4 GPa. The slopes of the low energy Raman modes at 288 cm^{-1} and 333 cm^{-1} change from 8.4 cm^{-1}/GPa to 2.2 cm^{-1}/GPa and from 2.7 cm^{-1}/GPa to 0.04 cm^{-1}/GPa, respectively. For the high frequency modes at 1429 cm^{-1} and 1442 cm^{-1}, the pressure coefficients change from 3.2 cm^{-1}/GPa to 6.2 cm^{-1}/GPa and from 7.0 cm^{-1}/GPa to 4.0 cm^{-1}/GPa, respectively. In addition, the band at 1559 cm^{-1} splits near ~4 GPa and the pressure coefficients of the split components are slightly different; namely, 4.9 cm^{-1}/GPa and 4.5 cm^{-1}/GPa for the higher and the lower energy component, respectively. The splitting of the mode at 1559 cm^{-1} and the changes in the pressure slopes of a number of modes, along with the reversibility of these effects upon pressure release after reaching 12 GPa, are the indication of a reversible structural phase transition that takes place near ~4 GPa. Concerning the changes taking place near ~15 GPa, they are apparently related to an irreversible transformation. More specifically, the rapid disappearance of the PP-mode in the pre-transitional pressure range, the drastic broadening of the Raman bands and the irreversibility behavior upon pressure release provide unambiguous evidence of an irreversible transformation to a highly disordered state.

The Raman data support the suggestion for a structural phase transition that may be related to minor changes in the packing of the linked linear polymeric chains. However, to prove that the singularities observed in the high-pressure Raman study are associated to a phase transition, an X-ray study of the structural aspects is necessary. A high-pressure X-ray powder diffraction study of the linear orthorhombic polymer of C_{60} has not revealed clear changes in the diffraction patterns near ~4 GPa [52]. In our opinion, this study is not conclusive since the low-resolution of the data along with the small number of peaks did not permit the complete refinement of the structure. The irreversible changes at ~15 GPa, from the well-resolved Raman spectrum of the photo-transformed 1D-O polymer to a diffuse one, is typical for a transition to a disordered phase. The rapid decrease of the $A_g(2)$ PP-mode intensity and the enhancement of the neighboring $H_g(8)$ and $G_g(6)$ modes, in the pre-transitional pressure regime, are reminiscent to the analogous behavior exhibited by these modes in the 2D-T and 2D-R polymeric phases of C_{60} before their further polymerization under high pressure [30], [42], [43]. Although pressure decreases preferentially the distance between the linear polymeric chains rather than the intercage distance within a chain, the bond formation between the chains is not always preferable. Namely, the C_{60} molecular cages belonging to adjacent polymeric chains may not have optimal relative orientations for the formation of new covalent bonds. Therefore, we expect that the new bonds be formed in a random way due to some distortion in the molecular orientations after photo-polymerization. As a result, the new high-pressure phase exhibits a high degree of disorder characterized by a random polymerization. Similarly, in the case of 2D-R polymer, the diffuse Raman spectrum of the high-pressure phase is also related to a disordered polymeric phase of C_{60} characterized by random covalent bonding between molecules belonging to adjacent 2D-polymeric planes of the initial rhombohedral phase. It is worth noticing that this behavior differs significantly from that of the 2D-T polymeric phase of C_{60}, in which the pressure-induced shortening of the intermolecular distances, accompanied by the optimal orientation of molecules, leads to a high degree of regularity in the formation of the out-of-plane covalent bonds.

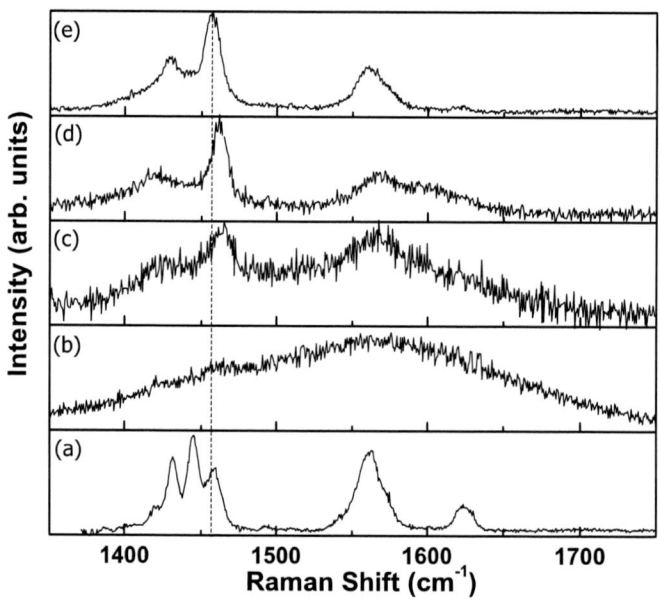

Figure 16. Raman spectra of various phases of the 1D-O polymer of C_{60} in the frequency region 1350-1750 cm^{-1} recorded at normal conditions. (a) The photo-transformed 1D-O polymer. (b) The recovered high-pressure phase immediately after pressure release. (c,d) The Raman spectra of the recovered high-pressure phase at different sites of the specimen recorded ~30 hours after pressure release. (e) The Raman spectrum of the pristine 1D-O polymer. Vertical line shows the position of the PP-mode of the pristine 1D-O polymer.

Thus, the decrease of the PP-mode intensity in the pre-transitional pressure range in linear and planar crystalline polymers of C_{60} is related to the destruction of a considerable number of double C = C bonds. The subsequent creation of numerous covalent links between molecules belonging to adjacent polymeric planes or chains results in regular or random cross-link polymerization of the initial polymers.

The recovered sample after pressure release was tested by means of micro-Raman probing in order to check its stability at ambient conditions. Figure 16 shows the Raman spectra of various phases of the 1D-O polymer of C_{60} in the frequency region 1350-1750 cm^{-1}, where the changes in the Raman response are more pronounced. The Raman spectrum of the photo-transformed 1D-O polymeric phase shown in Figure 16(a) differs significantly from that of the pristine 1D-O polymer shown in Figure 16(e). The Raman spectrum of the high-pressure phase of the photo-transformed 1D-O polymer immediately after pressure release is shown in Figure 16(b), whereas the in Figures 16(c) and 16(d) are shown the Raman spectra of the high-pressure phase at different sites of the sample ~30 hours after pressure release. These spectra indicate that the high-pressure phase is metastable and transforms rather quickly to a phase which demonstrates Raman features that resemble those of the initial 1D-O polymer of C_{60} {Figure 16(e)}. Nevertheless, the spectrum in Figure 16(d) is characteristic of a mixture of monomer and dimer forms of C_{60} as it follows from the frequencies of the $A_g(2)$ PP-mode which is shifted to higher energy with respect to that in the initial 1D-O polymer (indicated by vertical dashed line). It is worth noticing that this behavior is similar to that exhibited also by the 2D-T polymeric phase of C_{60} [42], [43]. The transformation of the recovered 1D-O material was observed at normal conditions without any special heat treatment of the sample except that due to the excitation beam during the Raman probing. The behavior of the recovered high-pressure phase of the photo-transformed 1D-O polymer differs from that of the high-pressure phase of the 2D-R polymer, which is more stable and transforms to a mixture of pristine and dimerized C_{60} only after sample annealing [30].

6. PHOTOLUMINESCENCE OF THE C60 POLYMERS AT HIGH PRESSURE

The optical transitions to the lowest excited singlet state of the C_{60} molecule (symmetry $^1T_{1g}$) are dipole forbidden, whereas the first allowed transition to $^1T_{1u}$ state has noticeably higher energy [53]. The fluorescence of the C_{60} molecule is related to the vibronically assisted electron-phonon transitions. As a result, the quantum yield of the fluorescence is relatively small: about 10^{-5} in solution and 7×10^{-4} in solid, slightly higher due to the presence of impurities and defects [54], [55]. As it follows from the calculations of the electronic structure and absorption spectra measurements the solid C_{60} is a direct-gap semiconductor with gap value in the range 1.5 - 1.8 eV [56], [57]. The PL spectrum of the single crystal of pristine C_{60} at low temperature is well-resolved, exhibiting fine structure associated with optical transitions from shallow defect levels, the so-called X-traps [58]. The polymerization of C_{60} leads to drastic changes in the crystal structure and the phonon spectrum of the C_{60} monomer [4], [5].

The electron energy spectra of the polymeric phases differ considerably from that of the pristine material, this has been predicted in a number of numerical calculations of electronic structure of the planar polymers of C_{60} [56], [59]-[61]. The decrease of the intermolecular distances, the deformation of the fullerene molecule cage and the lowering of molecular symmetry in polymeric phases of C_{60} affect significantly the electron energy spectrum of the pristine C_{60}. Numerical calculations, performed using the local-density approximation, predict that the tetragonal and rhombohedral polymeric phases of C_{60} are indirect low gap semiconductors and their electronic structure differs significantly from that of pristine C_{60} [53], [59]-[61]. The optical study of the C_{60} polymers has revealed some important changes in the PL spectrum of pristine C_{60} related to its polymerization, while the specimens identified as orthorhombic and tetragonal polymers were mainly a mixture of various planar polymeric phases of C_{60} [62]. In this paragraph, we present an overview of the results of the PL spectra measurements of perfect crystalline samples of the 1D-O, 2D-R and 2D-T polymeric phases of C_{60}, at normal conditions and as a function of pressure up to 4.0 GPa. The comparison of the PL spectra of the linear and planar polymers with those of the pristine C_{60}, as well as with the results of numerical calculations of the electronic structure, provide an insight in the changes associated with the polymerization of C_{60}.

The PL spectra of the C_{60} single crystals, 1D-O, 2D-R and 2D-T polymers of C_{60} are shown in Figure17. The PL spectrum of C_{60} single crystals, at normal conditions, consists of two broad peaks. The diffuse spectrum becomes well resolved when recorded at liquid helium temperature. The optical transitions from shallow defect levels located near the bottom of the conduction band show narrow bands at low temperatures [58]. The measurements at low temperature and high pressure have also revealed a band related with the transitions of the free excitons. The intensity of this band increases with pressure and its pressure coefficient is noticeably higher than those related to the localized excitons [63]. The PL spectrum of the linear 1D-O polymer is also diffuse but its structure differs from that of the pristine C_{60} in both the number of peaks and the intensity distribution among them. The planar polymers of C_{60} show additional changes in their PL spectra in comparison to the corresponding pristine C_{60} and 1D-O polymer recorded at normal conditions [64].

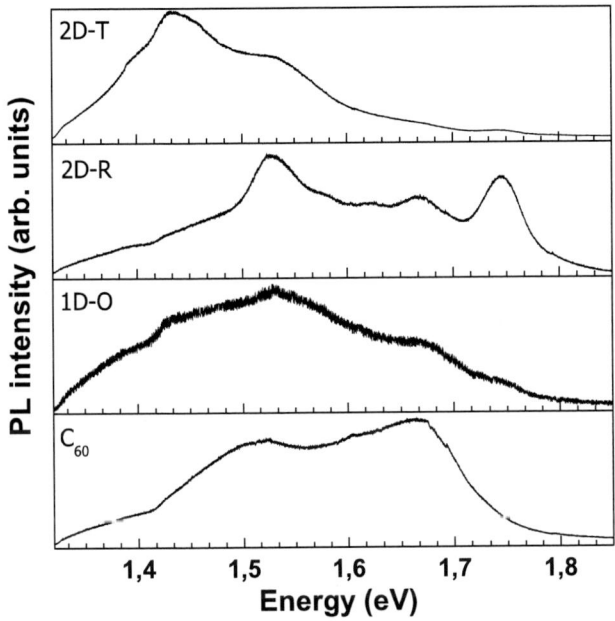

Figure 17. Photoluminescence (PL) spectra of the pristine C_{60}, 1D-O, 2D-R and 2D-T polymeric phases of C_{60} at ambient conditions.

The differences refer to the onsets of the spectra, their structure, and the number of bands as well as to their intensity distribution. The PL spectrum of the 2D-T polymer consists of two main bands similar to that of the pristine C_{60}, but its onset is shifted to lower energy by ~0.14 eV and the intensity distribution between these bands is reversed. In addition, the PL spectrum of the 2D-T polymer contains two very weak bands at the higher energy, which coincide with bands observed in the 2D-R polymer, therefore may be attributed to minute impurity of this phase in the 2D-T phase as we shall discuss below. On the contrary, the onset of the PL spectrum of the 2D-R polymer is slightly higher than that of the pristine C_{60}, while the spectrum is better resolved even at room temperature and contains four relatively narrow intense bands. Note, that the quality of the PL spectra is better than those reported earlier [62] that is related, in our opinion, to higher purity and better crystallinity of the polymeric specimens used in the present study.

The important difference between the PL spectra of the two planar polymers to that of the pristine C_{60} refers to their pressure behavior. Figure18 shows the PL spectra of the 2D-R polymer at various pressures up to ~3.5 GPa for increasing (left panel) and decreasing (right panel) pressure cycles.

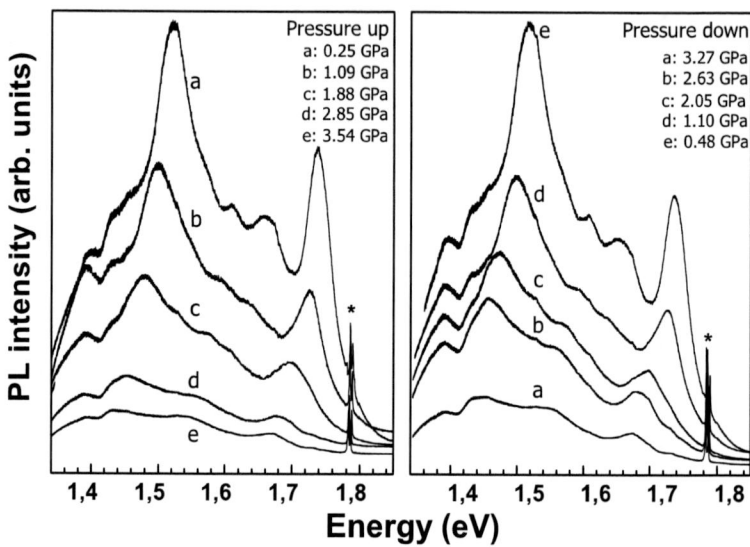

Figure 18. PL spectra of the 2D-R polymer at various pressures and room temperature for upstroke (left panel) and downstroke (right panel) pressure cycles. The ruby peaks are marked by asterisk.

As the pressure increases, the bands gradually shift to lower energies and the intensity of the PL spectrum decreases. The red pressure shift in the optical spectra of the fullerene-based materials is typical for solids with van der Waals intermolecular interaction [57], [63]. The pressure coefficients for the various bands of the 2D-R polymer vary in the region -0.022 eV/GPa to -0.028 eV/GPa. For comparison, the pressure coefficient of the free exciton band in the PL spectrum of the pristine C_{60} is about -0.086 eV/GPa [63]. The intensity distribution in the PL spectra is shown as measured: it is not corrected for the spectral response of the CCD. Thus, the rapid decrease of the PL intensity near the low energy part of the spectrum, below 1.4 eV, is related to the spectral cut-off of the CCD. On the other hand, the pressure-induced decrease of the PL intensity at the high-energy part of the spectrum, common characteristic for the pristine C_{60} and its polymeric phases, has another origin. It is associated, in our opinion, with the decrease of the fluorescence quantum yield, which is related to the pressure-induced enhancement of the singlet-triplet conversion. The proximity of the singlet and triplet electronic states in the C_{60} molecule [53] results in the relatively high phosphorescence of the fullerene that reduces the quantum yield of fluorescence. The energy gap between the singlet and triplet electronic states in the molecular solids decreases with the increase of pressure. This results in the enhancement of the singlet-triplet conversion and the increase of the phosphorescence intensity.

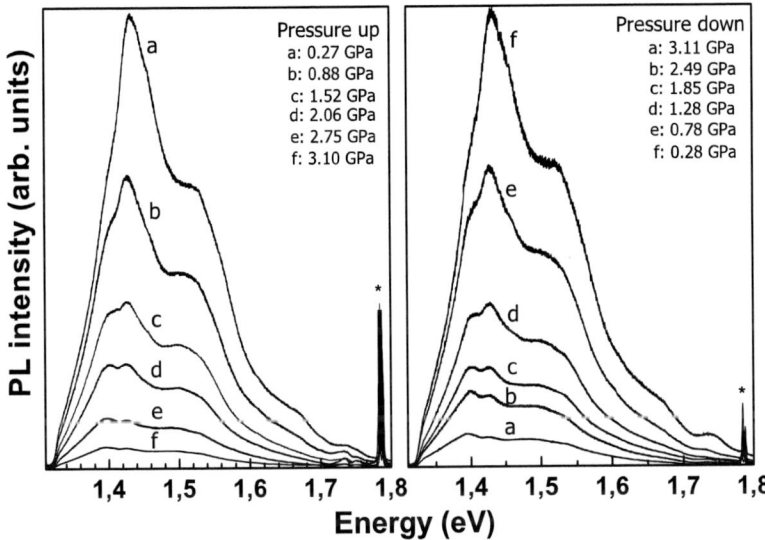

Figure 19. PL spectra of the 2D-T polymer at various pressures and room temperature for upstroke (left panel) and downstroke (right panel) pressure cycles. The ruby peaks are marked by asterisk.

The PL spectra of the 2D-T polymer at pressure up to 3 GPa for the increasing (left panel) and decreasing (right panel) pressure cycles are shown in Figure19. The pressure behavior of the PL spectra is similar, in general, to that of the 2D-R polymer, but the pressure coefficients for the main bands of the 2D-T polymer are smaller and vary between -0.013 eV/GPa and -0.015 eV/GPa. Note that the pressure behavior of the PL spectra for both planar polymers is reversible and all details of the spectra are restored after pressure release. The PL spectra of the 1D-O polymer for increasing (left panel) and decreasing (right panel) pressure cycles are shown in Figure 20. The intriguing feature, in the pressure behavior of the 1D-O polymer, is the irreversibility of the transformation in the PL spectra during the increasing-decreasing pressure cycle. The structure of the PL spectra after pressure release differs from that of the initial material. The difference is most pronounced at the high-energy part of the PL spectrum, which resembles, after pressure release, somehow the spectrum of the 2D-R polymer. This behavior is in agreement with the data, reported above in the high-pressure Raman study of the 1D-O polymer that demonstrates the irreversible photo-induced pressure-enhanced transformation associated with conjugation of the adjacent polymeric chains [47].

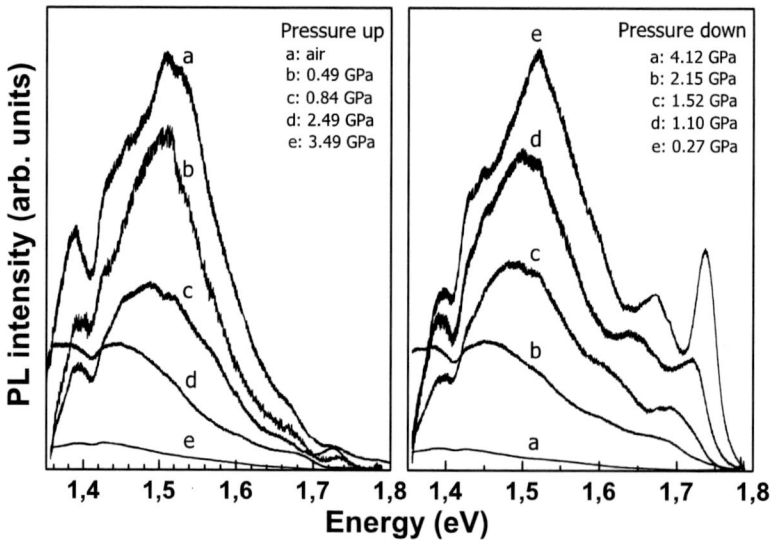

Figure 20. PL spectra of the 1D-O polymer at various pressures and room temperature for upstroke (left panel) and downstroke (right panel) pressure cycles. The ruby peaks are marked by asterisk.

The pressure dependence of the band positions in the PL spectra of the 2D-T (left panel) and 2D-R (right panel) polymers is shown in Figure 21. The open (closed) circles refer to the increasing (decreasing) pressure cycles, respectively, while the stars show the pressure dependence of the free exciton band of the pristine C_{60}. The solid and dashed lines are linear fittings to the experimental data related to the bands of the polymers and pristine C_{60}, respectively. Note that the slopes of the pressure dependencies differ for almost all bands of the two planar polymers except for the high-energy band. The initial position and the pressure dependence of this band coincide in the spectra of the 2D-R and 2D-T polymers, while its pressure coefficient $\partial E/\partial P = -0.022$ eV/GPa, being the same for the two polymers, differs from those of the other two bands of the 2D-T polymer. This means that the specimens of the 2D-T polymer contain small inclusions of the rhombohedral polymer, which result in the appearance of the weak first band at $E = 1.747$ eV in the PL spectra. The fact, that the preliminary Raman characterization of the samples does not show the presence of the other polymeric phase, means that the PL measurements are more sensitive to the purity of the specimens.

The characteristics of the PL spectra of the planar polymeric phases of C_{60} and the data related to the pressure dependence of the PL bands are strong indications that the electronic structure in these polymeric phases is different with respect to the pristine C_{60}. Numerical calculations of the electronic structure predict that the planar 2D-R and 2D-T polymers are indirect gap semiconductors with gap values of 0.35 eV and 0.72 eV for the rhombohedral and tetragonal polymeric phases, respectively [60], [61]. On the contrary, pristine C_{60} is a direct-gap semiconductor with gap energy ~1.5 eV as it follows from the local density approximation (LDA) numerical calculations [56]. The calculated gap values are noticeably smaller than the experimental values determined from the PL spectra of the polymeric phases. This discrepancy tends to be smaller for the pristine C_{60} in which the difference between the calculated and the experimental determination are smaller than those corresponding to the polymers. Furthermore, the most intriguing aspect in the PL behavior is that the onset of the PL spectrum of the 2D-R polymer is shifted to higher energy with respect to the position of the free exciton band of the pristine C_{60}. Normally the electronic spectra of the polymers should be shifted to lower energies: this is related to the decrease of the intermolecular distances in polymeric phases in comparison to those of the pristine C_{60} under high pressure. In fact, the situation is more complicated due to the deformation of the fullerene molecular cage, which can affect significantly the electronic structure of the polymer. The numerical LDA calculations take into account the deformed fullerene molecule cage and the reduced intermolecular distances in the polymeric phases, nevertheless the experimental results of the present work are rather far apart from the calculated data. A possible reason for this discrepancy may be related, in particular, with the structure of the experimental PL spectra of polymers, which represent mainly the direct phonon-assisted electronic transitions due to their relatively large intensity with respect to the indirect ones.

The important feature in the pressure behavior of the PL spectra is the noticeable difference in the pressure coefficients of the 2D-R and 2D-T polymers despite the fact that their bulk moduli B_0 are very close (the values of B_0 are 14.4, 28.1 and 29.9 GPa for the pristine C_{60}, 2D-R and 2D-T phases, respectively) [6], [40]. According to the X-ray diffraction measurements the compressibility of the 2D-R polymer within the polymeric sheets is almost three times larger than that of the 2D-T polymer, whereas the out-of-plane compressibility is almost the same [40]. The large difference in the pressure coefficients of the two polymers may be related to the difference in the in-plane compressibility of the polymers. Summarizing this paragraph, significant differences between the PL spectra of the linear 1D-O polymer, 2D-R, 2D-T planar polymers and solid C_{60} were found. These differences along with the difference in their pressure behavior provide the experimental proof that the polymerization of C_{60} results in considerable changes of the electronic structure of the pristine C_{60}. In addition, the experimental data regarding the energy gap values are qualitatively compatible with the results of numerical calculations, however they are rather far from quantitative agreement.

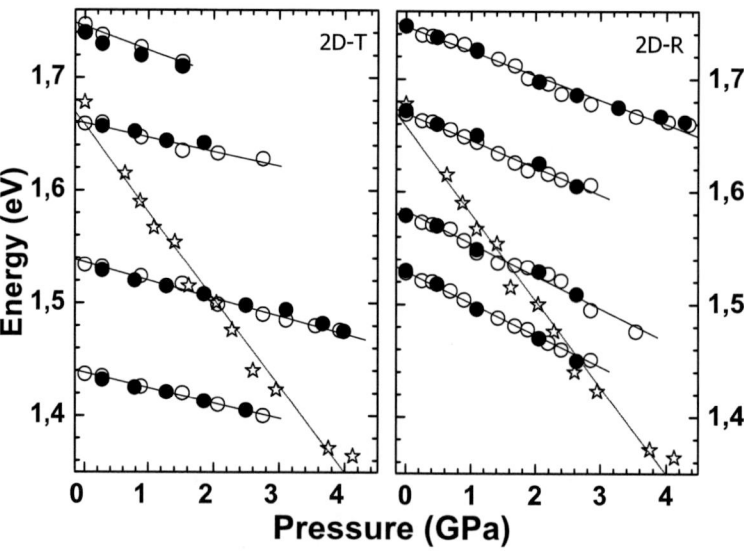

Figure 21. The pressure dependence of the band energies in the PL spectra of pristine C_{60} (stars), the 2D-T (left panel) and the 2D-R (right panel) polymers (circles). Open (closed) symbols represent data recorded for increasing (decreasing) pressure runs.

7. THERMAL STABILITY AND DECOMPOSITION KINETICS OF THE 2D-R POLYMER OF C_{60}

The stability of the fullerene polymers at ambient and elevated temperatures was studied by differential scanning calorimetry (DSC) and IR spectroscopy. These studies have shown that fullerene polymers are not stable at elevated temperature; heating to ~560 K results in the destruction of intermolecular C-C bonds and reversion to the initial C_{60} monomer phase [65],

[66]. The DSC studies performed for various polymers under a heating rate of 10-20 K/min show a strong endothermic peak between 525 and 565 K that was not observed during the cooling scan. The temperature of the irreversible transition depends on the polymeric phase and somewhat on the scanning rate, indicating that the polymer decomposition process is controlled by kinetics. The change of enthalpy related to the complete decomposition of polymers is the highest for the C_{60} dimers, its value decreases for the linear polymeric chains and becomes smallest for the planar polymeric phases [65]. The differences in the enthalpy change between the various polymeric phases and the kinetics of the polymer decomposition suggest the possible formation of intermediate polymeric/oligomeric states during the process of the temperature-induced decomposition of the polymeric fullerene networks. To study the polymer decomposition process and intermediate states formed, we have measured the spatially-resolved Raman spectra of the single crystalline 2D-R polymer of C_{60} after its treatment at various temperatures up to 600 K. The distinct difference between the Raman band frequencies of the various polymeric/oligomeric phases, namely the PP-mode position, allows their identification in the intermediate state that may appear upon the polymer decomposition. Raman spectra of the 2D-R polymer measured in the region of the $A_g(2)$ mode, after sample treatment for 0.5 hour at various temperatures, are presented in Figure 22. The spectra were measured at room temperature to avoid the sample damage that was observed within the laser spot for $T \geq 430$ K even at laser intensity as low as 0.005 mW.

The Raman spectra show that the 2D-R polymer is stable up to ~510 K, while at higher temperature a material transformation takes place as it can be inferred from the decrease of the PP-mode intensity, the increase of the background and the appearance of new Raman peaks. The transformation takes place through an intermediate state; the Raman spectrum of the material treated at 523 K has a relatively large background and new peaks typical for the PP-mode of the 2D-T and the 1D-O polymer as well as of the C_{60} dimers and monomers, which coexist with the PP-mode of the initial 2D-R polymer. Inclusion of 2D-T-like oligomers in the intermediate state can be deduced from the Raman line at ~1447 cm^{-1}, characteristic of the PP-mode in the 2D-T polymer. Their presence is related to the dissociation of 4 intermolecular C-C bonds of the initial 2D-R polymer, resulting in the creation of 2D-T-like oligomers having 8 intermolecular C-C bonds per C_{60} molecule. The Raman peak at ~1459 cm^{-1}, typical for linear polymeric chains, suggests the presence of inclusions related to the dissociation of 8 intermolecular C-C bonds of the initial 2D-R polymer and the formation of 1D-O-like oligomers having 4 intermolecular C-C bonds per C_{60} molecule.

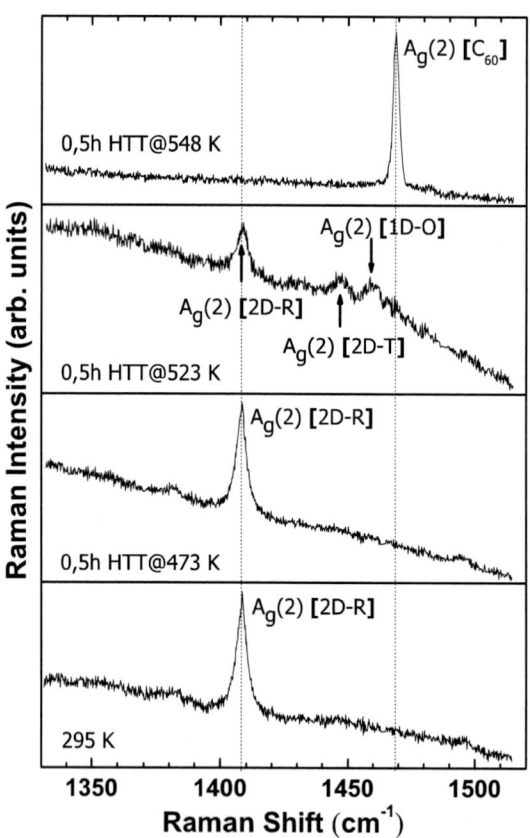

Figure 22. Raman spectra of the 2D-R polymer measured at room temperature in the frequency region of the $A_g(2)$ mode (left vertical line) after HTT for 0.5 hour at various temperatures. The arrows mark PP-modes related to oligomers in partially decomposed intermediate state. After complete decomposition at 548 K only the PP-mode of monomer C_{60} remains in the spectrum (right vertical line).

The intermediate state of the partially decomposed 2D-R polymer was observed up to 560 K where the material changes drastically its composition resulting in the domination of C_{60} monomers with some inclusion of C_{60} dimers. Figure 23 shows the relative intensities of the $A_g(2)$ PP-mode as a function of the treatment temperature of the initial state of the 2D-R polymer (open circles), the final state of the material after high-temperature treatment

(HTT) comprising of C_{60} monomers and dimers (closed circles) as well as the intermediate state comprising of 2D-T- and 1D-O-like oligomers (diamonds). The Raman spectra were measured after a number of 0.5 hour heat treatments at various temperatures on pieces of fresh sample. For most of the treatments, two or more spectra were averaged from different sample sites and/or sample pieces. The relative intensities of the various components, defined as the intensity of the $A_g(2)$ peak of each component normalized to their sum, reflect their relative concentration in the intermediate state. Thus, the concentration of the 2D-T- and 1D-O-like oligomers in the intermediate state increases with the increase of the treatment temperature up to its maximum at ~525 K, while at higher treatment temperature it gradually decreases to zero at ~560 K.

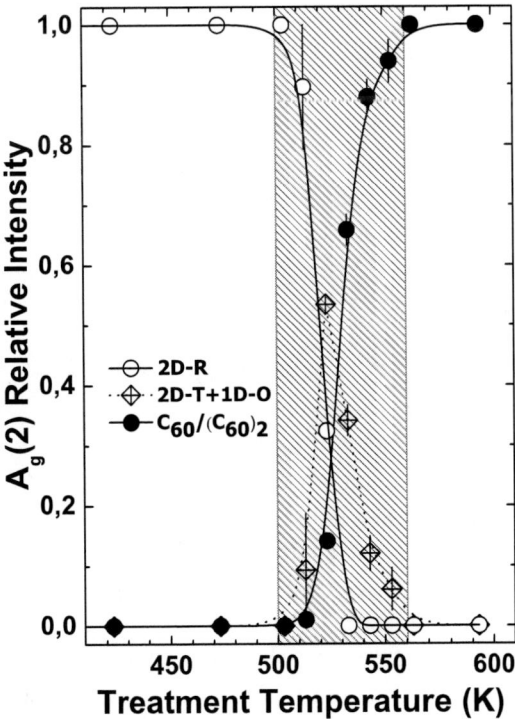

Figure 23. Intensity of the $A_g(2)$ peak attributed to the initial 2D-R polymer (open circles), the intermediate 2D-T and 1D-O oligomers (diamonds) and the C_{60} monomers/dimers (solid circles) normalized to their sum as a function of the treatment temperature for HTT lasted for 0.5 hour on pieces of fresh sample. The spectra were averaged from different sample sites and/or sample pieces, while the error bars refer to the standard error of the mean. The shaded area denotes the temperature region of the polymer decomposition; the lines through the data are guides to the eye.

The existence of 2D-T- and 1D-O-like oligomers in the intermediate state underlines the fact that the 12 intermolecular C-C bonds of the 2D-R polymer do not break simultaneously, in spite of their equivalence related to the hexagonal symmetry of the planar rhombohedral polymeric sheets. The broad lineshape of the Raman peaks related to the $A_g(2)$ mode of the 2D-T- and the 1D-O-like oligomers indicate either the structural disorder or the small size of the oligomer inclusions. It should be also noted, that the formation of oligomers was observed earlier in the IR spectra of the partially decomposed 2D-R polymer, while the two-step decomposition of the 2D-R polymer through the formation of an intermediate triangular cyclic trimer state was proposed to account for the observations [67]. According to our Raman data, the situation is more complicated; the polymer decomposition takes place through the formation of an intermediate state that is in fact a mixture of the initial 2D-R polymer, 2D-T- and 1D-O-like oligomers of C_{60} as well as monomer and dimers of C_{60}. Nevertheless, the long time HTT results to the complete decomposition of the material and the final state of the decomposed 2D-R polymer is pure monomer C_{60}, possibly with small inclusions of C_{60} dimers. This final state of the decomposed 2D-R polymer is structurally ordered as its Raman spectrum is characterized by relatively low background and sharp peaks. Moreover, no sign of amorphous carbon formation was evident in the Raman spectra for the temperature and treatment time used in our study.

To study the kinetics of the 2D-R polymer decomposition, the room temperature Raman spectra were measured for samples treated at various temperatures and heat treatment times. The Raman spectra of the 2D-R polymer in the frequency region of the $A_g(2)$ PP-mode after treatment at 513 K for 0.5-2.0 hours are illustrated in Figure 24. The treatment for 0.5 hour results in the decrease of the $A_g(2)$ mode intensity of the 2D-R polymer. The HTT for 1 hour results in further decrease of the $A_g(2)$ peak intensity and the appearance of $A_g(2)$ mode peaks associated with the presence of the 2D-T- and the 1D-O-like oligomers as well as with the monomeric C_{60}. Finally, the heat treatment for 2 hours leads to the practically complete decomposition of the 2D-R polymer as the $A_g(2)$ mode peak, attributed to the C_{60} monomer, dominates the Raman spectrum. Note that, the effect of heating has an additive character as there is no significant difference between the spectra recorded from a sample after continuous HTT at 513 K for a certain treatment time from those of a sample treated at the same temperature for the same total time but with intermediate cooling(s) to room temperature.

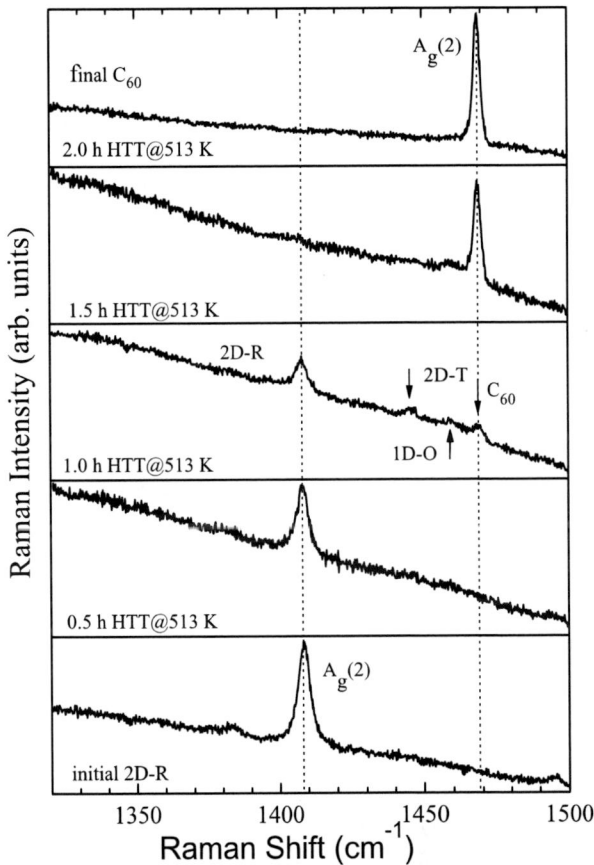

Figure 24. Raman spectra of the 2D-R polymer measured at room temperature in the high frequency region after HTT at 513 K for 0.5-2.0 hours. Arrows indicate the PP-modes of the intermediate 2D-T-like and 1D-O-like oligomers as well as of the monomer C_{60}. Vertical lines indicate the PP-mode of the initial 2D-R polymer and the final monomer C_{60}.

Figure 25 shows the intensity of the $A_g(2)$ mode of the 2D-R polymer normalized to the sum of the corresponding peaks associated with all polymeric/oligomeric phases of C_{60} as a function of the HTT time for different temperatures. The time required for the complete decomposition of the 2D-R polymer at 533 K is ~0.5 hour, while it increases at lower temperatures. The occurrence of polymer decomposition at temperatures as low as 503 K and the increase of the time needed for complete decomposition with decreasing treatment temperature, indicate the activation-type behavior of the polymer decomposition transition that is typical for chemical reactions. According to

DSC measurements, the minimum of the total energy related to the *fcc* monomeric state of C_{60} is higher by ~0.13 eV/molecule than the total energy minimum related to the 2D-R polymer of C_{60} [65].

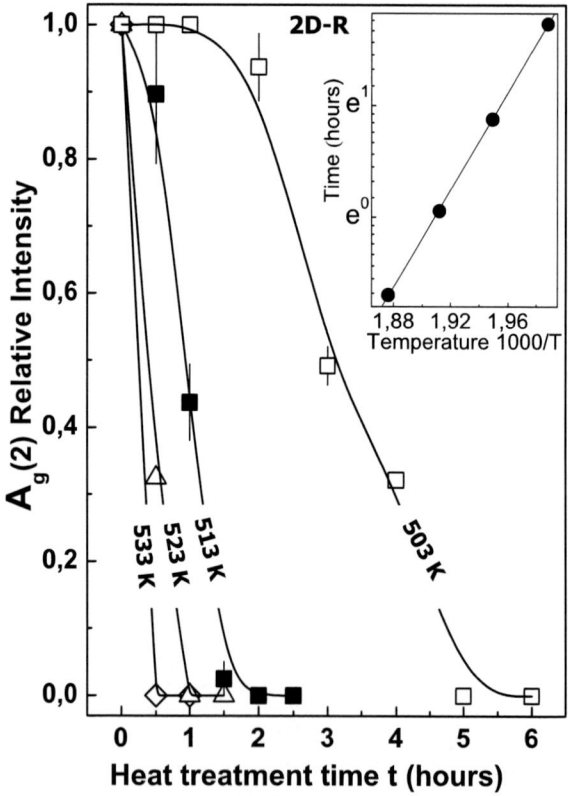

Figure 25. Intensity of the $A_g(2)$ peak of the initial 2D-R polymer normalized to the sum of the $A_g(2)$ peaks associated with the 2D-R polymer, 2D-T-like and 1D-O-like oligomers as well as with the monomer/dimer mixture of C_{60}, as a function of the heat treatment time at various temperature. Diamonds, triangles, circles and squares correspond to treatment temperatures of 533, 523, 513 and 503 K, respectively. The data were averaged from different samples and the error bars refer to the standard error of the mean, while the lines are guides to the eye. Inset: Arrhenius plot of the polymer's complete decomposition time versus the treatment temperature.

The stability of the C_{60} monomer at ambient conditions is due to the energy barrier that separates the polymeric state from the monomer state. The theoretically predicted energy barrier for the 2D-R polymer is 1.6 eV/molecule [59]. Another theoretical calculation of the energy barrier yields a value of 1.7

eV/molecule [68]. We can estimate the barrier value using our Raman data taking into account that the polymer decomposition time versus heat treatment temperature can be described by an Arrhenius equation:

$$\tau(T) = A \times \exp(E_A/k_B T),$$

where E_A is the activation energy (in fact the energy barrier between the polymer and monomer states), k_B is the Boltzmann constant, T is the treatment temperature and $\tau(T)$ is the temperature-dependent polymer decomposition time constant. The constant A, measured in time units, is related to the characteristic phonon frequency [69]. The Arrhenius plot of the polymer decomposition time versus the treatment temperature is shown in the inset of Figure 25. The experimental points exhibit a remarkably good linear dependence in logarithmic scale, yielding an activation energy of $E_A = 1.76 \pm 0.07$ eV/molecule. This value is smaller than the value of 1.9 ± 0.2 eV/molecule obtained in [69] from thermal expansion measurements on 2D-R polymeric powders. The activation energy obtained from the Raman experiments presented here is in good agreement with the calculated barrier energy; it is ~10% larger than the data reported in [59] and coincides within experimental accuracy with the data reported in [68]. It is important to note, that the enthalpy of the polymer decomposition extracted from DSC measurements, is not related to the energy barrier between the polymeric and the monomeric C_{60} state, but to the difference between the total energy minima. For example, the transition enthalpy for the 2D-R polymer obtained in [65] and [66] is equivalent to an energy difference between the two minima of ~0.13 eV/molecule and ~0.11 eV/molecule, respectively. We can estimate the complete decomposition time of the 2D-R polymer from the obtained data; at 300 K it is equal to ~4.49×10^8 years, but it collapses to ~62 hours at T=473 K. Thus, the fullerene polymers demonstrate rather large fragility at elevated temperatures that may be a reason preventing their practical use.

CONCLUSION

The crystalline polymers of C_{60} are not stable with respect to the high-pressure and high temperature treatment. The planar polymers show further polymerization at high pressure and room temperature. The Raman data provide a strong indication that the 2D-T polymer of C_{60} undergoes an irreversible phase transition above 20 GPa. The transformation takes place via a highly disordered pre-transitional state extending in a pressure range of ~4 GPa and having a rather diffuse Raman spectrum. The prominent Raman peaks of the high-pressure phase associated with the C_{60} molecular cage, as well as the irreversibility of the observed transformation, support the assumption of a further pressure-induced polymerization resulting in ordered 3D-cross-linked structure. This observation was confirmed in the X-ray study of the 2D-T polymer at pressure up to 37 GPa that revealed a transition at ~24 GPa associated with the formation of interlayer 3+3 cycloaddition along the body diagonal. The high-pressure 3D-polymeric phase demonstrates very high hardness; the bulk modulus determined from X-ray data is 407 GPa which is slightly smaller that that of the diamond (443 GPa).

The pressure behavior of the Raman modes the 2D-R polymer of C_{60} provides strong indication that the polymer undergoes an irreversible transformation at P~15 GPa. The initial well resolved Raman spectrum of the 2D-R polymer transforms to a diffuse one that is typical of a disordered state. Unlike to the 2D-T polymer, the high-pressure phase in 2D-R polymer is formed by the random creation of covalent bonds between adjacent polymeric layers due to various distortions in the molecular orientations.

In contradiction to the planar polymers of C_{60}, the linear orthorhombic polymer of C_{60} is not stable with respect to laser irradiation and transforms to a new polymeric phase. The laser intensity necessary for the photo-induced polymerization at ambient conditions is two orders of magnitude higher than that needed for the polymerization of the C_{60} monomer. The application of high pressure drastically increases the photo-polymerization rate and the transformation becomes almost instantaneous. The structural aspect of the photo-polymerization of the 1D-O polymer is related to conjugation of the adjacent polymeric chains. The high-pressure Raman study of the photo-transformed 1D-O polymer of C_{60} shows a possible structural phase transition at ~4 GPa, whereas further increase of pressure to ~15 GPa causes a transformation to a high-pressure phase with a diffuse spectrum typical of a disordered phase. As in the case of the 2D-R polymer, the high-pressure phase in photo-transformed 1D-O polymer is formed by the random creation of covalent bonds between polymeric layers due to some distortion in the molecular orientation. The high-pressure phases of linear and planar crystalline polymers of C_{60} recovered to ambient conditions are metastable and transform to a mixture of pristine and dimerized C_{60} under heating. This fact confirms the retention of the fullerene molecular cage in the high-pressure phases, while the similarity of the pre-transitional behavior of Raman spectra in linear and planar polymeric phases of C_{60} shows the common nature of the high-pressure phases related to their 3D cross-linked polymerization.

The crystalline linear and planar polymers of C_{60} show a gradual decomposition at high temperature. The 2D-R polymer composition changes drastically after treatment at ~560 K, resulting in the domination of C_{60} monomers with some inclusion of C_{60} dimers. The decomposition takes place through an intermediate state that, in fact, is a mixture of the initial 2D-R polymer, 2D-T- and 1D-O-like oligomers of C_{60} as well as monomer and dimer C_{60}. The final state of the decomposed 2D-R polymer is pure monomer C_{60} with, possibly, small inclusions of C_{60} dimers. The decomposition of the 2D-R polymer may occur at relatively low temperatures. The polymer decomposition time dependence on the treatment temperature is of the Arrhenius type while the energy barrier between 2D-R polymer and C_{60} monomer is 1.76±0.07 eV/molecule. The estimation of the complete decomposition time of the 2D-R polymer at 300 K gives ~1.19×10^8 years while at T=473 K it collapses to ~62 hours. Thus, the fullerene polymers demonstrate rather large fragility at elevated temperatures that may be a reason preventing their practical use.

ACKNOWLEDGMENTS

The support of the Russian foundation for basic research, (Russia) and the General Secretariat for Research and Technology, (Greece) is gratefully acknowledged. The authors would like to thank Drs. J. Arvanitidis, D. Christofilos, K. Papagelis, S. Assimopoulos, A. Soldatov, T. Wägberg, S. Yamanaka, V. Davydov and V. Agafonov for their help with the preparation and characterization of various polymers of C_{60}, and Profs. Y. Iwasa, B. Sundqvist, K. Prassides and S. Ves for helpful collaboration.

REFERENCES

[1] Rao, A. M.; Zhou, P.; Wang, K.-A.; Hager, G. T.; Holden, J. M.; Wang, Y.; Lee, W.-T.; Bi, X.-X.; Eklund, P. C.; Cornett, D. S.; Duncan, M. A.; Amster, I. J. *Science* 1993, 259, 955-957.

[2] Winter, J.; Kuzmany, H. *Solid. State Commun.* 1992, 84, 935-938.

[3] Stephens, P. W.; Bortel, G.; Faigel, G.; Tegze, M.; Janossy, A.; Pekker, S.; Oszlanyi, G.; Forro, L. *Nature* 1994, 370, 636-639.
[4] Iwasa, Y.; Arima, T.; Fleming, R. M.; Siegrist, T.; Zhou, O.; Haddon, R. C.; Rothberg, L. J.; Lyons, K. B.; Carter, H. L.; Hebard, A. F.; Tycko, R.; Dabbagh, G.; Krajewski, J. J.; Thomas, G. A.; Yagi, T. *Science* 1994, 264, 1570-1572.
[5] Nunez-Regueiro, M. ; Marques, L.; Hodeau, J.-L. ; Bethoux, O.; Perroux, M. *Phys. Rev. Lett.* 1995,74, 278-281.
[6] Sundqvist, B. *Advances in Physics* 1999, 48, 1-134.
[7] Brazhkin, V. V.; Lyapin, A. G.; Popova, S. V. *JETP Lett.* 1996, 64, 802-807.
[8] Blank, V.; Popov, M.; Buga, S.; Davydov, V.; Denisov, V. N.; Ivlev, A. N.; Marvin, B. N.; Agafonov, V.; Ceolin, R.; Szwarc, H.; Rassat, A. *Phys. Lett. A* 1994, 188, 281-286.
[9] Blank, V. D.; Buga, S. G.; Dubitsky, G. A.; Serebryanaya, N. R.; Popov M. Yu.; Sundqvist, B. *Carbon* 1998, 36, 319-343.
[10] Marques, L.; Mezouar, M.; Hodeau, J.-L.; Nunez-Regueiro, M.; Serebryanaya, N. R.; Ivdenko, V. A; Blank, V. D.; Dubitsky, G. A. *Science* 1999, 283, 1720-1723.
[11] Chernozatonskii, L. A.; Serebryanaya, N. R.; Mavrin, B. N. *Chem. Phys. Lett.* 2000, 316, 199 -204.
[12] Moret, R. *Acta Cryst. A* 2005, 61, 62-76.
[13] Dzyabchenko, A. V.; Agafonov, V.; Davydov, V. A. *Crystallogr. Rep.* 1999, 44, 18-24.
[14] Ruoff, R. S.; Ruoff, A. L. *Nature* 1991, 350, 663-664.
[15] Okada, S.; Saito, S.; Oshiyama, A. *Phys. Rev. Lett.* 1999, 83, 1986-1989.
[16] Burgos, E.; Halac, E.; Weht, R.; Bonadeo, H.; Artacho, E.; Ordejon, P. *Phys. Rev. Lett.* 2000, 85, 2328-2331.
[17] Davydov, V. A.; Kashevarova, L. S.; Rakhmanina, A. V.; Senyavin, V. M.; Ceolin, R.; Szwarz, H.; Allouchi, H.; Agafonov, V. *Phys. Rev. B* 2000, 61, 11936-11945.
[18] Arvanitidis, J.; Papagelis, K.; Meletov, K. P.; Kourouklis, G. A.; Ves, S.; Kordatos, K.; Wudl, F.; Prassides, K. *Phys. Rev. B* 1999, 59, 3180-3183.
[19] Kourouklis, G. A.; Ves, S.; Meletov, K. P. *Physica B* 1999, 265, 214-222.
[20] Wägberg, T.; Soldatov, A.; Sundqvist, B. *Eur. Phys. J. B* 2006, 49, 59-65.

[21] Senyavin, V. M.; Davydov, V. A.; Kashevarova, L. S.; Rakhmanina, A. V.; Agafonov, V.; Allouchi, H.; Ceolin, R.; Sagon, G.; Szwarc, H. *Chem. Phys. Lett.* 1999, 313, 421-425.
[22] Jayaraman, A. *Rev. Sci. Instrum.* 1986, 57, 1013-1031.
[23] Barnett, D.; Block, S.; Piermarini, G. *J. Rev. Sci. Instrum.* 1973, 44, 1-9.
[24] Davydov, V. A.; Kashevarova, L. S.; Rakhmanina, A. V.; Agafonov, V.; Allouchi, H.; Ceolin, R.; Dzyabchenko, A. V.; Senyavin, V. M.; Szwarc, H. *Phys. Rev. B* 1998, 58, 14786-14790.
[25] Winter, J.; Kuzmany, H.; Soldatov, A.; Persson, P.-A.; Jacobsson, P.; Sundqvist, B. *Phys. Rev. B* 1996, 54, 17486-17492.
[26] Persson, P.-A.; Edlund, U.; Jacobsson, P.; Johnels, D.; Soldatov, A.; Sundqvist, B. *Chem. Phys. Lett.* 1996, 258, 540-546.
[27] Porezag, D.; Pedersson, M. R.; Frauenheim, T.; Kohler, T. *Phys. Rev. B* 1995, 52, 14963-14970.
[28] Solin, S. A. ; Ramdas, A. K. *Phys. Rev. B* 1970, 1, 1687-1698.
[29] Arvanitidis, J.; Meletov, K. P.; Papagelis, K.; Ves, S.; Kourouklis, G. A.; Soldatov, A.; Prassides, K. *J. Chem. Phys.* 2001, 114, 9099-9104.
[30] Meletov, K. P.; Arvanitidis, J.; Kourouklis, G. A.; Iwasa, Y.; Prassides, K. *Chem. Phys. Lett.* 2002, 357, 303-315.
[31] Haines, J.; Leger, J. M. *Solid. State Commun.* 1994, 90, 361-363.
[32] Goncharov, A. F.; Makarenko, I. N.; Stishov, S. M. *JETP Letters* 1985, 41, 184-187.
[33] Brazhkin, V. V.; Lyapin, A. G.; Popova, S. V.; Voloshin, R. N.; Antonov, Yu. V.; Lyapin, S. G.; Klyuev, Yu. A.; Naletov, A. M.; Mel'nik, N. N. *Phys. Rev. B* 1997, 56, 11465-11471.
[34] Brazhkin, V. V.; Lyapin, A. G.; Popova, S. V.; Klyuev, Yu. A.; Naletov, A. M. *J. Appl. Phys.* 1998, 84, 219-226.
[35] Weiler, M.; Sattel, S.; Giessen,T.; Jung, K.; Ehrhardt, H.; Veerasamy, V. S.; Robertson, J. *Phys. Rev. B* 1996, 53, 1594-1608.
[36] Nemanich, R. J.; Solin, S. A. *Phys. Rev. B* 1979, 20, 392-401.
[37] Shroder, R. E.; Nemanich, R. J.; Glass, J. T. *Phys. Rev. B* 1990, 41, 3738-3745.
[38] Long, D. A. *Raman Spectroscopy*, McGraw-Hill: London, UK, 1976, p. 158-159.
[39] Dam Hieu Chi; Iwasa, Y.; Takano, T.; Watanuki, T.; Ohishi, Y.; Yamanaka, S. *Phys. Rev. B* 2003, 68, 153402/1-153402/4.
[40] Kawasaki, S.; Yao, A.; Matsuoka, Y.; Komiyama, S.; Okino, F.; Touhara, H.; Suito, K. *Solid. State Commun.* 2003 125, 637-640, (2003).

[41] Leger, J. M.; Haines, J.; Davydov, V. A.; Agafonov, V. *Solid. State Commun.* 2002, 121, 241-244.
[42] Meletov, K. P.; Arvanitidis, J.; Tsilika, I.; Assimopoulos, S.; Kourouklis, G. A.; Ves, S.; Soldatov, A.; Prassides, K. *Phys. Rev. B* 2001, 63, 054106/1-054106/4.
[43] Meletov, K. P.; Tsilika, I.; Assimopoulos, S.; Arvanitidis, J.; Kourouklis, G. A.; Ves, S.; Sundqvist, B.; Wägberg, T. *Chem. Phys. Lett.* 2001, 341, 435-441.
[44] Martin, M. C.; Du, X.; Kwon, J.; Mihaly, L. *Phys. Rev. B* 1994, 50, 173-183.
[45] Meletov, K. P.; Arvanitidis, J.; Assimopoulos, S.; Kourouklis, G. A.; Sundqvist, B. *JETP Lett.* 2002, 95, 736-747.
[46] Kourouklis, G. A.; Meletov, K. P. *New Diamond and Frontier Carbon Technology* 2002, 12, 303-314.
[47] Meletov, K. P.; Davydov, V.A.; Rakhmanina, A. V.; Agafonov, V.; Kourouklis, G. A. *Chem. Phys. Lett.* 2005, 416, 220-224.
[48] Bethune, D. S.; Meijer, G.; Tang, W. C.; Rosen, H. J.; Golden, W. G.; Seki, H.; Brown, C. A.; de Vries, M. S. *Chem. Phys. Lett.* 1991, 179, 181-186.
[49] Fagerstrom, J.; Stafstrom, S. *Phys. Rev. B* 1996, 53, 13150-13158.
[50] Davydov, V. A.; Kashevarova, L. S.; Rakhmanina, A. V.; Agafonov, V.; Ceolin, R.; Szwarc, H. *JETP Lett.* 1998, 68, 928-933.
[51] Ciabini, L.; Santoro, M.; Bini, R.; Schettino, V. *Phys. Rev. Lett.* 2002, 88, 085505/1-085505/4.
[52] Le Parc, R.; Levelut, C.; Haines, J.; Davydov, V.A.; Rakhmanina, A. V.; Papoular R. J.; Belova, E. E.; Chernozatonskii, L. A.; Allouchi, H.; Agafonov, V. *Chem. Phys. Lett.* 2007, 438, 63-66.
[53] Negri, F.; Orlandi, G.; Zerbetto, F. *Chem. Phys. Lett.* 1998, 144, 31-37.
[54] Wang, Y. *J. Phys. Chem.* 1992, 96, 764-767.
[55] Lane, P. A.; Swanson, L. S.; Ni, Q.-X.; Shinar, J.; Engel, J. P.; Barton, T. J.; Jones, L. *Phys. Rev. Lett.* 1992, 68, 887- 890.
[56] Saito, S.; Oshiyama, A. *Phys. Rev. Lett.* 1991, 66, 2637-2640.
[57] Meletov, K. P.; Dolganov, V. K.; Zharikov, O. V.; Kremenskaya, I. N.; Ossip'yan, Yu. A. *J. Phys. France* 1992, 2, 2097-2105.
[58] Guss, W.; Feldman, J.; Göbel, E. O.; Taliani, C.; Mohn, H.; Müller, W.; Häussler, P.; ter Meer, H.–U. *Phys. Rev. Lett.* 1994, 72, 2644-2647.
[59] Hui, C.; G. Scuseria, E. *Phys. Rev. Lett.* 1995, 74, 274-277.
[60] Okada, S.; Saito, S. *Phys. Rev. B* 1997, 55, 4039-4041.
[61] Okada, S.; Saito, S. *Phys. Rev. B* 1999, 59, 1930-1936.

[62] Venkateswaran, U. D.; Sanzi, D. S.; Krishnappa, J.; Marques, L.; Hodeau, J.-L.; Nunes-Regueiro, M.; Rao, A. M.; Eklund, P. C. *Phys. Stat. Sol. (b)* 1996, 198, 545-552.
[63] Meletov, K. P.; Negrii, V. D. *JETP Lett.* 1998, 68, 248-252.
[64] Meletov, K. P.; Kourouklis, G. A. *Chem. Phys. Lett.* 2005, 403, 338-342.
[65] Iwasa, Y.; Tanoue, K.; Mitani, T.; Yagi, T. *Phys. Rev. B.* 1998, 58, 16374-16377.
[66] Korobov, M. V.; Senyavin, V. M.; Bogachev, A. G.; Stukalin, E. B,; Davydov, V. A.; Kashevarova, L. S.; Rakhmanina, A. V.; Agafonov, V.; Szwarc, H. *Chem. Phys. Lett.* 2003, 381, 410-415.
[67] Korobov, M. V.; Bogachev, A. G.; Popov, A. A.; Senyavin, V. M.; Stukalin, E. B.; Dzyabchenko, A. V.; Davydov, V. A.; Kashevarova, L. S.; Rakhmanina, A. V.; Agafonov, V. *Carbon* 2005, 43, 954-961.
[68] Saito, S.; Okada, S. *AIP Conf. Proc.* 1998, 442, 198-202.
[69] Nagel, P.; Pasler, V.; Lebedkin, S.; Soldatov, A.; Meingast, C.; Sundqvist, B.; Persson, P.-A.; Tanaka, T.; Komatsu, K.; Buga, S.; Inaba, A. *Phys. Rev. B* 1999, 60, 16920-16927.

Chapter 2

BIO-BASED (CO)POLYESTERS CONTAINING 1,4-CYCLOHEXYLENE UNITS: CORRELATIONS BETWEEN STEREOCHEMISTRY AND PHASE BEHAVIOR

Annamaria Celli[*], Paola Marchese, Simone Sullalti and Corrado Berti
Department of Civil, Environmental and Materials Engineering
University of Bologna, Via Terracini 28, 40131
Bologna, Italy

ABSTRACT

Today there is an increasing interest to develop novel eco-friendly polymers, i.e. materials produced from renewable resources, with low energy consumption, non-toxic to the environment, in some cases also biodegradable and with good mechanical performances. In the field of aliphatic polyesters, novel (co)polymers, containing 1,4-cyclohexylene units, are very promising materials, as obtainable from biomass, environmentally degradable and characterized by good mechanical properties. Moreover, in these polyesters the cyclic units have two possible configurations (cis and trans) which strongly influence the final phase behavior. Indeed, the trans isomer is more rigid and symmetric than

---

[*] E-mail address: annamaria.celli@unibo.it

the cis. Highly symmetrical units tend to improve the chain packing with a consequent increment in crystallinity and crystalline perfection. On the other hand, the cis isomer introduces kinks into the main chain, which hinder the formation of stable crystals. Thus, at high trans content the polyesters are characterized by a relative high degree of crystallinity, whereas at low trans content the polymers are amorphous. Therefore, accordingly to the final cis/trans ratio, the phase behavior of the homopolymers and copolymers significantly changes and the stereochemistry of the cycloaliphatic units result to be a key factor to tailor the final thermal properties of the material. In this paper the properties of some homopolymers and copolymers, containing the 1,4-cyclohexylene units with different cis/trans ratio, are discussed just in terms of the correlations between stereochemistry and phase behavior.

1. INTRODUCTION

In recent years there has been a great interest on environmentally and safety problems, which become more and more urgent in front of the amounts of not biodegradable waste and widespread pollution. As to the plastics, great difficulties are connected to the substitution of traditional polymers, which are generally prepared from non-renewable feedstock and are resistant to biodegradation. Biobased chemicals and sustainable materials, instead, are now a priority.

In this context all-aliphatic polyesters are excellent candidates to have a significant role in the next generation of plastics. Indeed, they often derive from renewable resources and potentially have some good properties, such as low toxicity and biodegradability. Moreover, with respect to the widely appreciated industrial thermoplastics, such as poly(butylene terephthalate) (PBT) and poly(ethylene terephthalate) (PET), aliphatic polyesters show improved capabilities, in terms of UV stability, as they do not contain any functional groups that absorb UV light, resistance to weather, heat and humidity, which make them ideal polymers in outdoor applications.

It is known, however, that the mechanical properties of aliphatic polyesters sometimes fail with respect to the properties of aliphatic-aromatic polyesters. Poly(R)-3-hydroxybutyrate (PHB), for example, is a brittle material and poly(alkylene dicarboxylate)s are characterized by very low melting peak temperatures (about ~ 40-80°C) [1], [2].

Introduction of alicyclic units to the main chain of the polymer can be a way to increase the rigidity of the macromolecular chains. Recently, for example, Liu and Turner [3] describe the preparation of a systematic series of random copolyesters using different cycloaliphatic diesters, which are, for the Authors, the most suitable monomers to achieve high Tg values, up to 155°C. Moreover, conformational transitions of cyclohexylene rings in the backbone originate secondary relaxations in dynamical mechanical spectrum, which contribute to improve the performances of the materials.

1,4-cyclohexane dicarboxylic acid (CHDA), dimethyl-1,4-cyclohexane dicarboxylate (DMCD), and 1,4-cyclohexane dimethanol (CHDM) (see Scheme 1) are the monomers commercially available to introduce the 1,4-cyclohexylene units in polyesters.

The synthesis and properties of polyesters and copolyesters containing these alicyclic moieties were studied at the beginning of the eighties by Eastman Chemical Company, interested to develop materials with excellent tensile strength, stiffness and impact properties as well as materials to be used as improved hot melt adhesives. After the first patents [4], [5], now there is a new interest for the polyesters deriving from cyclic diacid/diesters and diols or containing aliphatic C_6 rings, especially after the recent improvements in the preparation procedure [6]-[9].

In our opinion, today the use of the cyclic monomers of Scheme 1 to prepare all-aliphatic polyesters present further advantages. Although CHDA, DMCD, and CHDM are now obtained from petroleum resources, however, they can be prepared from bio-based terephthalic acid, starting from limonene and other terpenes [10].

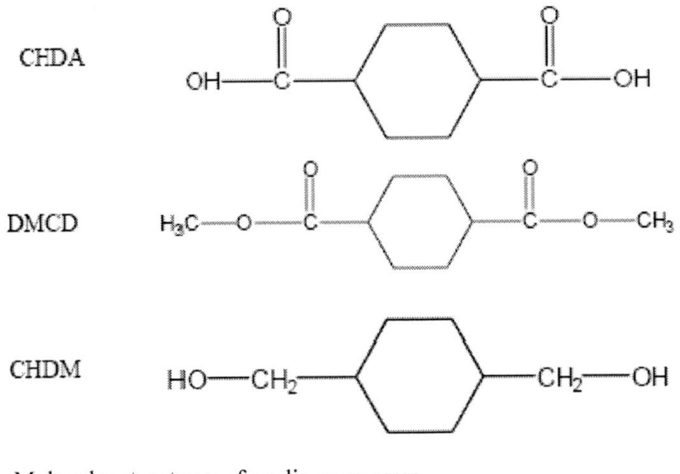

Scheme 1. Molecular structures of cyclic monomers.

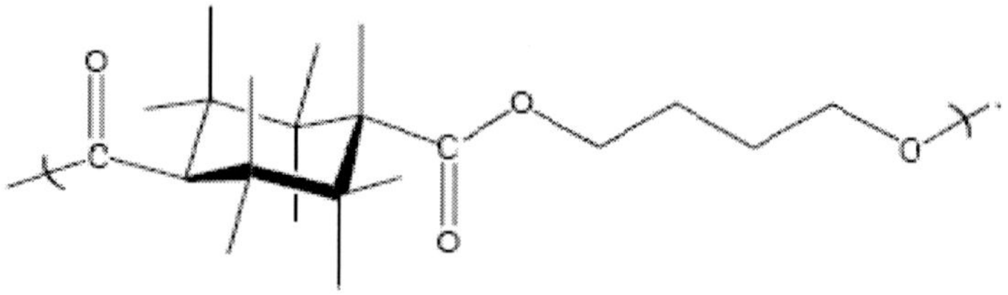

trans

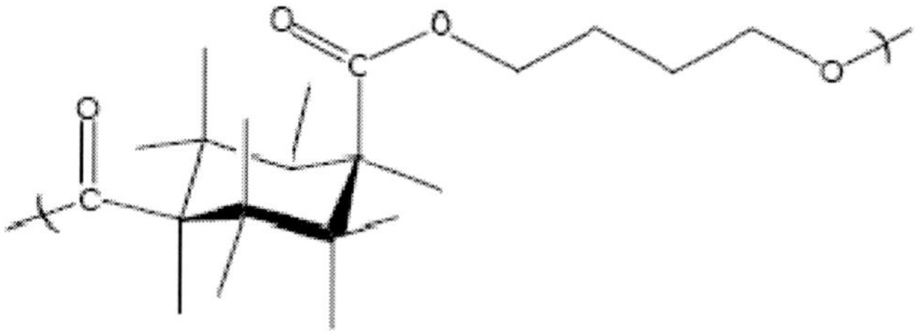

cis

Scheme 2. Trans and cis configurations of the monomeric unit in Poly(Butylene 1,4-Cyclohexanedicarboxylate) (PBCHD).

Therefore, polymers derived, for example, from DMCD and a diol obtainable from biomass (as 1,4-butanediol, which can be prepared, for example, from succinic acid [11]) can be considered fully sustainable materials. Moreover, we have observed that the presence of the 1,4-cyclohexylene units along a macromolecule does not hinder the attack of microorganisms in some homopolymers and copolymers [12]. Therefore, the polyesters containing the 1,4-cyclohexylene units can be considered environmentally degradable materials and are very promising, novel biopolyesters.

Again, it is notably that the 1,4-cyclohexylene unit shows another remarkable peculiarity: it can have two possible configurations, cis and trans, as described in Scheme 2 for the monomeric unit of the Poly(butylene 1,4-cyclohexanedicarboxylate) prepared from DMCD and 1,4-butanediol, called PBCHD. From the literature, which studied the incorporation of cycloaliphatic monomers in polyesters, polyamides, polyurethanes [13]-[18], and from some of our researches [19]-[21], it results that the isomeric ratio of the cycloaliphatic residues along the chains is the key factor which determines the phase behavior of the materials. This is a crucial point, which in our opinion has not been explored fully in the literature. Indeed, although there are numerous reports about the properties of various cyclic ester containing polyesters, there are few papers of all-aliphatic (co)polyesters based on a systematic variation of the stereochemistry of the cyclic monomers. For example, for CHDA in [22] only a cis/trans ratio of about 40/60 has been investigated, whereas in [23] and [24] only cis/trans ratios of 80/20 and 70/30, respectively, have been used. Instead, we believe it should be possible to obtain completely different properties of the copolymers using a different cis/trans ratio of the starting materials.

Therefore, this paper focus on the preparation and characterization of polyesters and copolyesters from a cyclic diester, DMCD, with a systematic variable cis/trans ratio, from about 80/20 to 0/100 mol%. The goal is a wide discussion on the correlations existing between the stereochemistry of the 1,4-cyclohexylene ring and the final phase behavior of the novel materials.

2. EXPERIMENTAL PART

2.1. Materials

Two commercial samples of dimethyl-1,4-cyclohexane dicarboxylate (DMCD), with 22% and 100% of trans isomer content, 1,4-butanediol (BD), 1,4-cyclohexane dimethanol (CHDM) with 66 mol% of trans isomer, 1,3-propanediol, dimethyl terephthalate (DMT), dimethyl adipate (DMA), titanium tetrabutoxide (TBT) (all from Aldrich chemicals) were high purity products, used as received. Scheme 1 describes the monomers containing an alicyclic unit.

Poly(buthylene terephthalate) (PBT) was a commercial product supplied by General Electric and was used as a reference material. Poly(propylene terephthalate) (PPT) was synthesized in our laboratory [25]. Ecoflex was a commercial product by BASF.

In the text the polyesters prepared from DMCD and BD are named with the code PBCHD-xx, (see Scheme 3) where xx indicates the percentage of the aliphatic rings, derived from DMCD, in trans configuration.

The polyesters derived from DMCD and CHDM are indicated with the code PCCD-Dxx-Eyy (see Scheme 3), where xx indicates the percentage of C6 rings deriving from CHDM (D) in trans configuration and yy the percentage of C6 rings deriving from DMCD (E) in trans configuration. The PCCD samples here considered are all prepared from CHDM containing 66 mol % of trans isomer.

The copolymers derived from BD, DMA, and DMCD are named (4-6)-co-PBCHDxx-a/b, where xx indicates the percentage of the aliphatic rings, derived from DMCD, in trans configuration and a/b is the feed molar ratio of the DMA/DMCD.

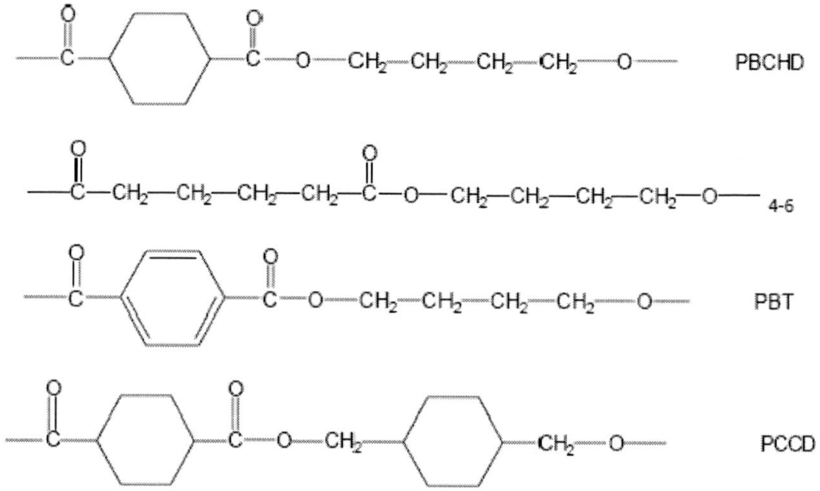

Scheme 3. Molecular structure of the repeating units of poly(butylene-1,4-cyclohexanedicarboxylate) (PBCHD), Poly(butylene adipate) (4-6), poly(butylene terephthalate) (PBT), and poly(1,4-cyclohexylenedimethylene 1,4-cyclohexanedicarboxylate) (PCCD).

2.2. Sample Preparation

All the syntheses are two stage polycondensations carried out in the presence of TBT as catalyst. The syntheses of PBCHD and PCCD samples are described in [19] and [20], respectively. For 4-6 the procedure of synthesis is similar to those discussed in [2] for analogous poly(alkylene dicarboxylate)s. As example, the syntheses of a PBCHD specimen and of a (4-6)-co-PBCHD copolymer are here reported.

2.2.1. Synthesis of Poly(Butylene 1,4-Cyclohexanedicarboxylate) (PBCHD)

DMCD (32,24 g, 0,171 mol), BD (20,5 g, 0.228 mol), TBT (0,02 g, 0,059 mmol) were placed into a round-bottomed wide-neck glass reactor (250 ml capacity). The reactor was closed with a three-neck flat flange lid equipped with a mechanical stirrer and a torque meter which gives an indication of the viscosity of the reaction melt. The reactor was immersed into a silicone oil bath preheated to 200°C. The first stage was conducted at atmospheric pressure under nitrogen atmosphere and the mixture was allowed to react for 120 min under stirring with continuous removal of water. The second stage was started by gradually reducing the pressure to 0.02 mbar while the temperature was raised to the final value of 220°C. These conditions were reached within 60 min, using a linear gradient of temperature and pressure, and maintained for 180 min.

2.2.2. Synthesis of (4-6)-Co-PBCHD90-50/50 Copolyester

DMA (16.20 g, 0.093 mol), DMCD 100% trans (16.22 g, 0.081 mol), DMCD 22% trans (2.40 g, 0.012 mol), BD (20.12 g, 0.223 mol), TBT (0.02 g, 0.047 mmol) were placed into a round bottom wide-neck glass reactor (250 ml capacity). The reactor was closed with a three-necked flat flange lid equipped with a mechanical stirrer and a torque meter which gives an indication of the viscosity of the reaction melt. The reactor was immersed into a silicone oil bath preheated to 200°C.

The first stage was conducted at atmospheric pressure under nitrogen atmosphere and the mixture was allowed to react for 90 min under stirring with continuous removal of water. The second stage was started by gradually reducing the pressure to 0.2 mbar while the temperature was raised to the final value of 220°C. These conditions were reached within 90 min, using a linear gradient of temperature and pressure, and maintained for 120 min.

The feed DMA/DMCD molar ratios used for the syntheses of the (4-6)-co-PBCHD copolymers are 30/70, 50/50, and 70/30 in order to obtain copolyesters with different compositions.

2.3. Characterization

The ^{1}H NMR spectra were recorded at room temperature on samples dissolved in CDCl3 using a Varian Mercury 400 spectrometer, the proton frequency being 400 MHz. Molecular weights (expressed in equivalent polystyrene) were determined by gel permeation chromatography (GPC), using a Hewlett Packard Series 1100 liquid chromatography instrument equipped with a PL gel 5µ Mixed-C column. Chloroform was used as eluent and a calibration plot was constructed with polystyrene standards. The thermogravimetric analysis (TGA) was performed using a Perkin-Elmer TGA7 thermobalance under nitrogen atmosphere (gas flow 40 ml/min) at $10°C·min^{-1}$ heating rate from 50°C to 900°C. The calorimetric analysis was carried out by means of a Perkin-Elmer DSC6, calibrated with high purity standards.

The measurements were performed under nitrogen flow. In order to cancel the previous thermal history, the samples (ca. 10 mg) were initially heated to different temperatures, varying from 160 to 260°C according to the sample characteristics, kept at high temperature for 1 min, and then cooled to a temperature range varying from -70 to 20°C at $10°C·min^{-1}$. After this thermal treatment, the samples were analyzed by heating to 160-260°C at $10°C·min^{-1}$ (2nd scan).

During the cooling scan the crystallization temperature (T_{CC}) and the enthalpy of crystallization (ΔH_{CC}) were measured. During the 2nd scan the glass transition temperature (Tg), the cold crystallization temperature (Tch) and enthalpy (Δ Hch), the melting temperature (Tm) and the enthalpy of fusion (Δ Hm) were measured. Tg was taken as the midpoint of the heat capacity increment associated with the glass-to-rubber transition.

Specimens for dynamic mechanical measurements were obtained by injection molding in a Mini Max Molder (Custom Scientific Instruments) equipped with a rectangular mold (30 x 8 x 1.6 mm3).

The molded samples were rapidly cooled in water and then dried in a oven at 50°C under vacuum overnight. Dynamic mechanical measurements were performed with a dynamic mechanical thermal analyzer (Rheometrics Scientific, DMTA IV), operated in the dual cantilever bending mode, at a frequency of 3 Hz and a heating rate of 3 $°C·min^{-1}$, over a temperature range from -150 to a final temperature varying from 100 to 150°C, according to the sample characteristics.

3. RESULTS AND DISCUSSION

3.1. Preparation and characterization of homopolymers

3.1.1 Synthesis

The synthesis of samples with different cis/trans ratio of the 1,4-cyclohexylene rings, derived from DMCD, is possible since the final cis/trans ratio depends on the isomeric content of the diester. Two DMCD monomers, characterized by 100 and 22 mol% of trans isomer respectively, are available on the market. An adequate physical mixture of these two components enables us to obtain the desired stereochemistry in the final polymer.

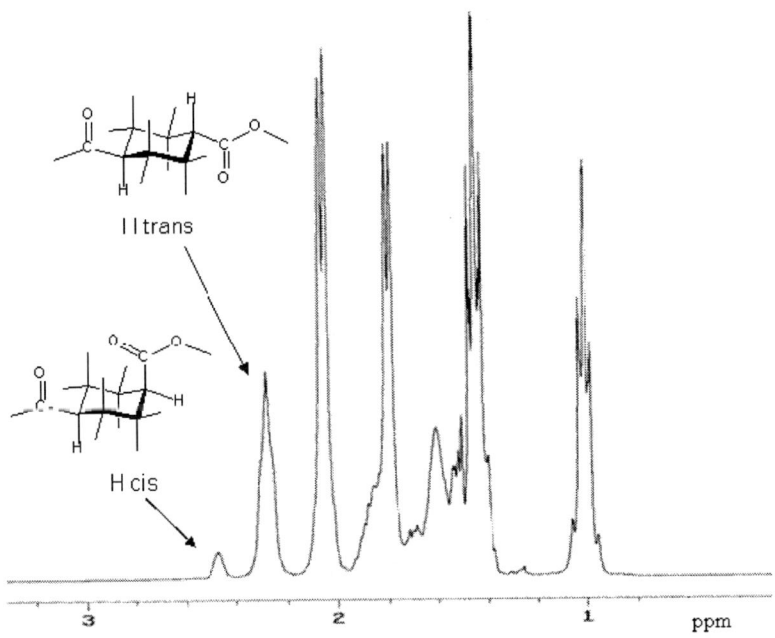

Figure 1. ^{1}H-NMR spectrum of PCCD-D66-E90 sample, with the indication of the signal used to calculate the cis/trans ratio.

Isomerization reactions, which can change the initial cis/trans content towards the thermodynamically stable 34/66 mol%, could take place. However, they are favored when the synthesis or thermal treatments are carried out at temperatures higher than 260°C and for longer than 1 h, and in the presence of acidic groups. Indeed, the use of 1,4-cyclohexane dicarboxylic acid (CHDA) as monomer gives rise to a 5-7% degree of isomerization, indicating a catalytic effect of the carboxylic acid towards isomerization [26]. These are the reasons why all the syntheses were performed by starting from DMCD (and not CHDA), and the temperatures did not exceed 220-240°C. Therefore, during the syntheses from DMCD, isomerisation is virtually absent. Moreover, for the syntheses of PCCD, we have verified that CHDM is unable to isomerise in the experimental conditions used.

Table 1. Molecular characteristics and thermal properties of PBCHD samples, with different cis/trans ratio, compared with those of a PBT specimen

Sample	trans %[a] in DMCD units of the polymer	Mw · 10^{-3} [b]	Mw/Mn[b]	T_{cc}[c] °C	ΔH_{cc}[c] J·g^{-1}	T_g[d] °C	T_m[d] °C	ΔH_m[d] J·g^{-1}
PBCHD-20	24	57.0	2.2	-	-	-12	-	-
PBCHD-50	52	88.6	2.8	-	-	-7	-	-
PBCHD-70	72	77.6	2.3	79	34	-2	122	22
PBCHD-80	80	78.4	2.3	106	37	1	132-141	27
PBCHD-90	91	54.9	2.3	130	45	6	150-158	37
PBCHD-100	100	73.4	2.5	149	48	10	165-171	47
4-6	-	90.0	2.5	32	67	-58	52-57	70
PBT	-	47.7	2.4	189	48	42[e]	224	43

[a] Calculated by 1H NMR
[b] Measured by GPC in CHCl3
[c] Measured in DSC during the cooling scan at 10°C·min^{-1}.
[d] Measured in DSC during the 2nd heating scan at 10°C·min^{-1}.
[e] From [28].

Table 2. Molecular characteristics and thermal properties of PCCD samples, with different cis/trans ratio

Sample	trans %[a] in DMCD units of the polymer	$M_w \cdot 10^{-3}$ [b]	M_w/M_n [b]	T_{cc} [c] °C	ΔH_{cc} [c] J·g^{-1}	T_g [d] °C	T_{ch} [d] °C	ΔH_{ch} [d] J·g^{-1}	T_m [d] °C	ΔH_m [d] J·g^{-1}
PCCD-D66-E20	24	43.3	2.2	-	-	38	-	-	-	-
PCCD-D66-E50	52	83.7	2.7	-	-	50	-	-	-	-
PCCD-D66-E70	66	71.7	2.2	-	-	55	142	3	181	4
PCCD-D66-E80	81	75.2	2.2	154	23	60	-	-	200 - 205	23
PCCD-D66-E90	91	56.9	2.3	188	29	65	-	-	218 -	30
PCCD-D66-E100	97	62.0	2.1	204	33	-	-	-	226 230	34

[a] Calculated by ^{1}H NMR
[b] Measured by GPC in CHCl$_3$
[c] Measured in DSC during the cooling scan at 10°C·min^{-1}.
[d] Measured in DSC during the 2nd heating scan at 10°C·min^{-1}.

The cis/trans isomeric ratio in polymers can be evaluated by ^{1}H NMR analysis. An example of an ^{1}H-NMR spectrum of PCCD is reported in Figure 1. The ratio of the areas of the signals centred at 2.3 (trans isomer) and 2.5 (cis isomer) ppm has been used to calculate the trans percentage [22], [27].

The final trans content varies from 24 to 100 mol% for both the series of PBCHD and PCCD samples, as reported in Table 1 and 2.

The molecular weight data, calculated by GPC, show that all the samples have significantly high and similar molecular weights, that is they are suitable for comparison in terms of thermal behaviour.

3.1.2. Effect of the Molecular Structure on Thermal Properties

The effect of the introduction of the 1,4-cyclohexylene moieties to an aliphatic macromolecular chain can be highlighted by comparing the thermal properties of the poly(butylene adipate), which is named 4-6, with those of PBCHD. Figure 2, for example, shows the DSC traces of 4-6 and PBCHD-100 and Table 1 reports the calorimetric data. It is noteworthy that both polymers are semicrystalline, with a high degree of crystallinity and a significant capability of crystallizing. Indeed, crystallization exotherms from the melt are very sharp and intense peaks. However, crystallization and melting temperatures of PBCHD-100 are shifted to about 100°C higher than those of 4-6. The Tg values are -58°C for 4-6 and vary from -12 to 10°C for all the PBCHDs, i.e. differ of about 50-70°C. All these data confirm that the macromolecular chains are characterized by a notably high rigidity in the presence of the alicyclic units [3]. Moreover, as it will be discussed in the following section, the chair (preferred) or boat conformations of the 1,4-cyclohexylene units are suitable for good chain packing with the formation of very stable and perfect crystals. However, this conclusion is valid only for the PBCHD samples at high trans percentage (see Section 3.1.3).

In any case, it is evident that the introduction of the 1,4-cyclohexylene units to the aliphatic macromolecules causes very interesting properties, such as relatively high melting temperatures and good chain rigidity.

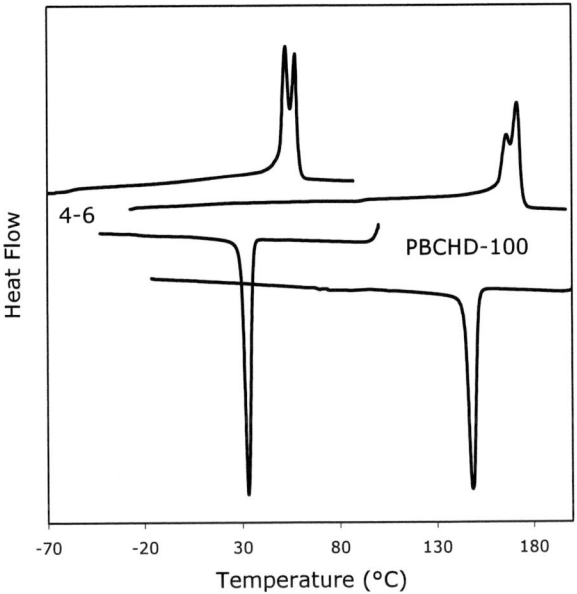

Figure 2. DSC thermograms for 4-6 and PBCHD-100 homopolymers.

In order to have a complete overview of the properties of alicyclic polyesters, it should be interesting to compare their thermal behavior with that of the traditional aliphatic-aromatic polyesters. Indeed, PBCHD is the all-aliphatic counterpart of the aliphatic-aromatic PBT (see Scheme 3). The difference in chemical structures is connected to the substitution of the 1,4-cyclohexylene group with the terephthalate unit of PBT. This difference causes lower T_g, T_C and T_m values in PBCHD than in PBT, as shown in Table 1 [28]. Indeed, in PBT, and in general in polyesters based on terephthalic acid, the coplanarity between the carbonyl and phenyl groups restricts the rotational angles about C_{phenyl}-CO to 0 and 180°, even if rotations about the terephthaloyl residue virtual bond, resulting in nonplanar conformations, are also probable, with a barrier which increases for rotation angles increasing from 0 to 90° [29], [30]. The planar conformation favors the molecular packing of the chains in the crystal, and enhances the attractive intermolecular interactions between the ester groups of neighboring chains. As a result, the aromatic polyesters exhibit high melting points [31]. The low flexibility of the chains, moreover, induces a high Tg value. On the other hand, in PBCHD the presence of the aliphatic ring, in chair or boat conformations, excludes the planarity of the CO-C6 system; moreover the cyclohexyl groups are conformationally more mobile than the rigid phenyl ring. For these reasons the melting temperatures and Tg values are considerably lower [24].

A similar behavior can be observed in Table 3 where two polyesters derived from 1,3-propanediol and DMCD (named PPCHD) are compared with a sample of poly(propylene terephthalate) (PPT) [25]. Also in this case, the molecular structures differ for aliphatic and aromatic rings. It is worth noting that the aliphatic rings inside the chains induce higher flexibility and, thus, notably lower Tg values. At the same time, the capability of crystallizing in PPCHD is significantly reduced with respect to PPT and the melting peak is more than 80°C lower.

Analogously, in [2] PET and the homologous, aliphatic poly(ethylene 1,4-cyclohexanedicarboxylate) (PECHD) are compared and some data collected in Table 3. It is confirmed that the presence of terephthalate units makes the chains more rigid and improve the crystallizability of the material.

Finally, it must be remarked that alicyclic polyesters and copolyesters, from DMCD, are also characterized by a high thermal stability.

Figure 3 shows some example of thermogravimetric curves for a sample of purified polylactic acid (PLA) (from Natureworks), a PBT and some polyesters and copolyesters containing the 1,4-cyclohexylene units. It is evident that the most stable materials are exactly these latter.

This result indicates that the substitution of the aromatic ring with an aliphatic ring improves the thermal stability of the materials. The same observations have been made, for example, by Wang et al. [23] for copolymers based on PET: the introduction of the 1,4-cyclohexylene rings causes an increment of the thermal stability of PET of 5-10 degrees, as a function of the aliphatic ring content. Moreover, also the substitution of the aliphatic sequence, in PBCHD, with another aliphatic ring, in PCCD, contribute to further stabilize the material.

In a similar way, the thermal stability of some aliphatic-aromatic random copolyesters increases as the cyclic units, obtained from CHDM, increase. [32].

Table 3. Molecular characteristics and thermal properties of PPCHDs, with different cis/trans ratio, and PECHD, compared with those of PPT and PET specimens

Sample	trans %[a] in DMCD units of the polymer	$M_w \cdot 10^{-3}$ [b]	M_w/M_n [b]	T_{ch} [c] °C	ΔH_{ch} [c] J·g^{-1}	T_g [c] °C	T_m [c] °C	ΔH_m [c] J·g^{-1}
PPCHD-50	53	62.4	2.4	-	-	-1	-	-
PPCHD-90	87	65.6	2.5	80	28	7	140	29
PPT[d]	-	65.0	2.3	-	-	49	223	68
PECHD[e]	60	55.3	2.3	-	-	14	-	-
PET[e]	-	34.3	2.4	-	-	82	252	37

[a] Calculated by ^{1}H NMR.
[b] Measured by GPC.
[c] Measured in DSC during the 2nd heating scan.
[d] From [25].
[e] From [22].

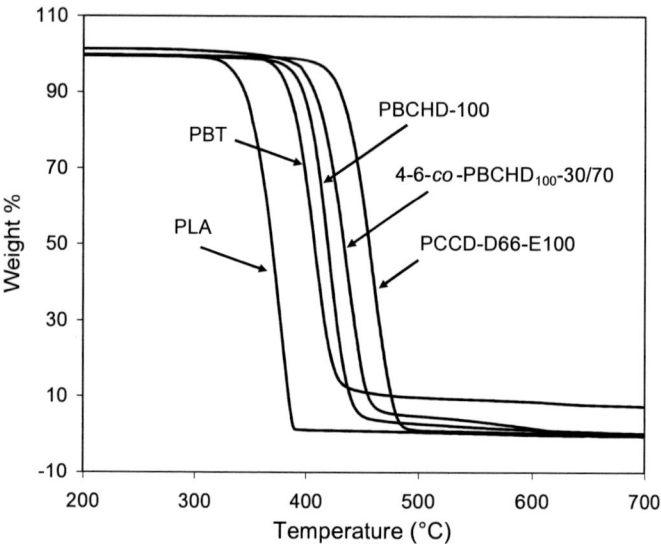

Figure 3. Thermogravimetric curves, obtained in nitrogen at 10°C·min⁻¹ for different aliphatic and aromatic polyesters and copolyesters.

3.1.3. Analysis of the Phase Behaviour for PBCHD and PCCD Homopolymers

The two isomers of the 1,4-cyclohexylene units, cis and trans (see Scheme 1), strongly influence the solid-state behaviour and produce in a polymers a wide variety of properties. Figure 4 and 5 show the thermal behavior of PBCHD samples with different cis/trans ratio and Table 1 reports the calorimetric data.

It must be remarked that only certain PBCHD samples crystallize during the cooling scan from the melt at 10°C·min⁻¹, in particular only PBCHD-70, -80, -90, and -100, i.e. only the samples with a trans content ≥ 70%. For these samples, the exothermal peaks become more and more intense and narrow as the trans content increases. Accordingly, T_{CC} and ΔH_{CC} have a notable increment.

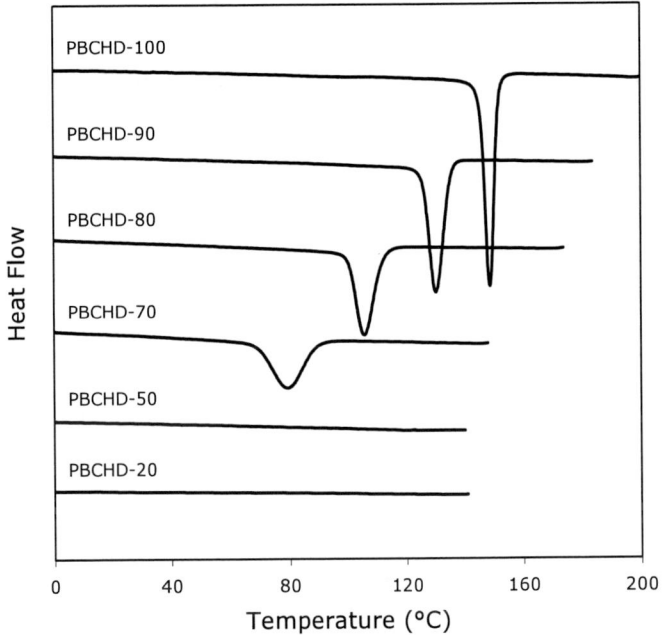

Figure 4. DSC thermograms of PBCHD samples during the cooling scan from the melt at 10°C·min^{-1}.

In particular, from PBCHD-70 to PBCHD-100 T_{CC} increases by about 70°C and the enthalpy doubles. These differences appear very significant and indicate that the trans configuration of the C6 ring along the chain improves the capacity of the samples to crystallize. On the other hand, it is significant that the PBCHD-50 and PBCHD-20, with the lowest percentages of trans stereoisomer, are not able to rearrange towards an ordered state at all.

The 2nd heating scans (see Figure 5) confirm that some PBCHD samples (-20 and -50) have the characteristics of completely amorphous materials. Indeed, they do not show any evidence of cold crystallization and following melting process during the heating scan in the experimental conditions used. On the other hand, the samples with a trans content higher than or equal to 70% are semicrystalline, with melting temperatures which increase considerably with the increment of the trans content, from 122°C for PBCHD-70 up to 165-170°C for PBCHD-100. Correspondently, ΔHm also reaches high values. For the melting process too, these variations are very significant and correspond to drastic changes in the final properties of the materials.

The tendency of the aliphatic rings in trans configuration to favour crystallizability can be explained by considering that in this configuration the polymeric chain assumes a "stretched" form and a high symmetry. These are conditions favourable to the chain packing. On the other hand, the cis isomer introduces kinks into the chain, which hinder the formation of stable crystals. Literature widely supports these observations. For example, in liquid crystalline polyesters only the trans isomer is compatible with melt anisotropy [33]. The cis units represent "kinks" which disrupt the rigid-rod main–chain structure, leading to non-ordered polymers [34], [35].

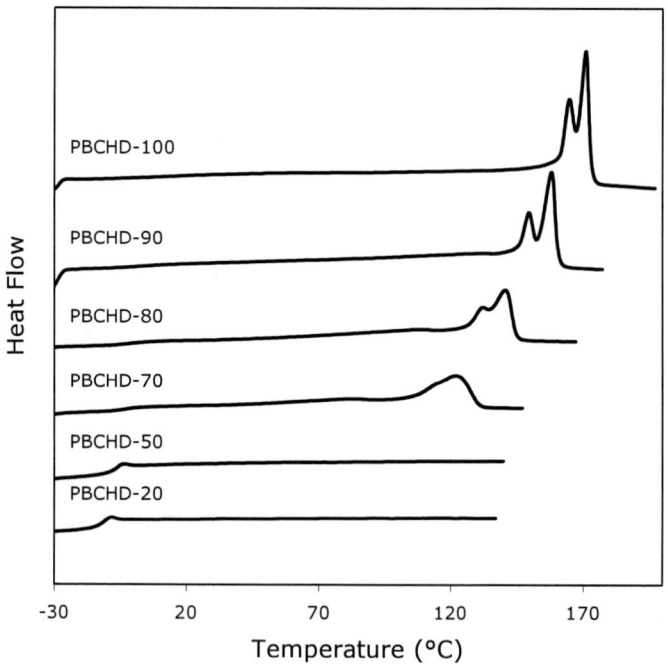

Figure 5. DSC thermograms of PBCHD samples during the 2nd heating scan at 10°C·min-1.

Moreover, in partially cycloaliphatic polyamides, characterized by the insertion of 1,4-cyclohexylene rings along the chains, from solid-state NMR studies and WAXD analyses the cis isomer is found only in the amorphous regions, whereas the trans isomer is located both in the crystalline and in the amorphous phases [16], [17]. This means that the crystals are formed only by the trans isomer, whereas the cis isomer is fully excluded from the crystals.

Again, Kricheldorf and Schwarz [13], who studied the thermodynamically controlled cis/trans ratio of polyesters deriving from CHDA, observed that the favoured trans content of the crystalline material is 100%. Moreover, the formation of stable crystals of trans isomer seems to be the driving force for the observed cis → trans isomerization in various polyesters [13], [36].

Therefore, it is confirmed that the stereochemistry of the 1,4-cyclohexylene ring plays a fundamental role in determining the characteristics of the phase behaviour of PBCHD.

A similar behaviour has been found in PCCD samples. From the data of Table 2 and from Figure 6 it is noteworthy that only certain samples crystallize during the cooling scan from the melt at $10°C \cdot min^{-1}$. In particular only D66-E80, D66-E90, and D66-E100 samples – in other words, only those samples with the highest trans content – crystallize. As the trans content increases, the temperature at which crystallisation from the melt occurs (T_{CC}) and enthalpy (ΔH_{CC}) tend also to increase. Therefore, the trans conformation of the C6 rings along the chain improve the capacity of the samples to crystallize, in terms of the crystallization rate, crystallinity and crystal perfection. It is significant that the other samples, with a lower percentage of trans stereoisomer, are totally unable to rearrange towards an ordered state.

Figure 7 shows the 2nd heating scans of the PCCD samples. It is confirmed that some PCCDs (D66-E20 and D66-E50) have the characteristics of completely amorphous materials. D66-E70 sample, instead, presents a certain capacity to reach a small amount of order (the enthalpies of crystallization and melting are about $4 \; J \cdot g^{-1}$).

As regards the glass transition temperature, Table 1 and 2 show the experimental data for PBCHDs and PCCDs, respectively. It is necessary to emphasize that for the samples with the highest level of crystallinity (PBCHD-100, PCCD-D66-E100) the determination of Tg is affected by a too high degree of uncertainty as the percentage of amorphous phase is limited. In any case, it is evident that the Tg values increase with the increment of the trans percentage. For example, from PBCHD-20 to PBCHD-100, Tg increases by about 30°C (from -12 to 10°C). For PCCDs Tg increases by about 25°C (from 38 to 65°C) with the increment of the trans content from 20 to 90%. Similarly, in polyamides obtained from CHDA and various diamines, Tg increases up to about 17°C when the trans content of the 1,4-cyclohexylene moieties changes from 23 to 100% [37].

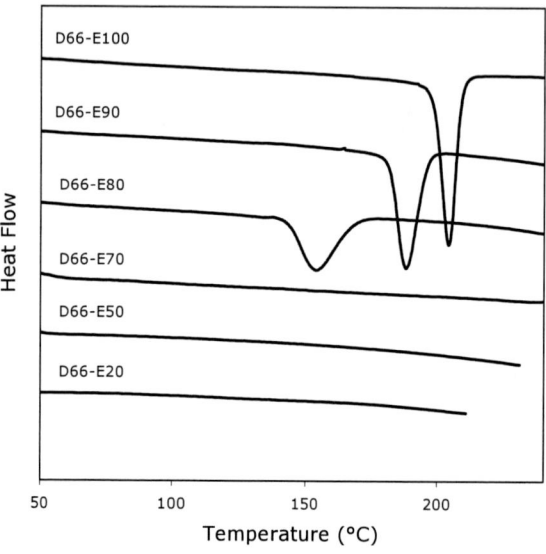

Figure 6. DSC thermograms of PCCD samples during the cooling scan from the melt at 10°C·min^{-1}.

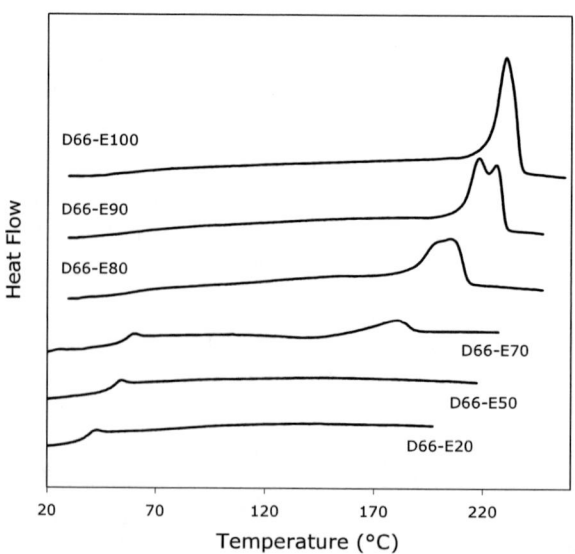

Figure 7. DSC thermograms of PCCD samples during the 2nd heating scan at 10°C·min^{-1}.

In this case, to explain the effect of the stereochemistry of the cycloaliphatic ring on the glass transition, it is necessary to recall that the most important factors influencing the Tg values are chain flexibility, symmetry, and steric hindrance and bulkiness of the side groups attached to the backbone chain. In amorphous polymers it has been observed that an increment of symmetry can induce an increment of Tg, as seen for example in 1,4 ring substituted nylon 66 copolyamides which have higher Tg values with respect to 1,2 and 1,3 ring substituted copolyamides [38]. Indeed, in 1,4 ring polymers a better molecular fit is achieved in the polymer backbone, resulting in a better chain packing and improved orientation, which would restrict the movement of the chains upon heating. The same consideration can be made for PBCHD and PCCD, where the trans isomer is more symmetrical than the cis. Moreover, in the presence of high trans % the high Tg values observed can be attributed also to the notable level of crystallinity which creates numerous impediments to the chain mobility in the amorphous state.

3.1.4. Relationship between Molecular Structure and Phase Behavior

3.1.4.1. PBCHD

In order to emphasize the effect of the trans isomer content on the thermal properties of our aliphatic polyesters, a statistical approach has been used.

The chemical structure of PBCHD has been analyzed by considering that the 1,4-cyclohexanedicarbonyl moiety is the unit whose characteristics (in terms of cis (c) or trans (t) isomer) have a great influence on the thermal transitions typical of the crystalline state. By focusing the attention on the sequence of three monomeric units, i.e. the sequence of three C6 rings, eight different triads are possible, based on the configuration of the rings: ttt, ctt, tct, ttc, cct, tcc, ctc, ccc. As the polymer chain is very long, it can be considered as formed by a high number of triads and each triad can assume one of the eight configurations described above, in a random distribution. Therefore, the probability (P) of finding a particular triad, for example the probability Ptct of finding the tct triad, within the chain, is calculated from the product of the probability (f_c or f_t) of finding the cis or trans configuration for each ring:

$$Ptct = (f_t \cdot f_c \cdot f_t) \cdot 100, \quad (1)$$

f_t and f_c are calculated by the trans and cis content of the DMCD unit in the polymer, reported in Table 1, divided by 100. Table 4 shows the probabilities of finding the eight possible triads within the chains in the PBCHD samples which show the capacity to crystallize.

The correlations between the T_{CC} and Tm values for the semicrystalline PBCHD samples and the probabilities of finding the above-mentioned triads are shown in Figure 8 and 9, respectively. For PBCHD-80, -90 and -100, which show a double melting peak, only the temperature of the first melting peak is reported, due to the fact that it corresponds to the melting of the crystals actually formed at T_{CC} during the cooling scan. Indeed, the second melting peak was attributed to melting-recrystallization and remelting processes occurring during the calorimetric scan.

Firstly, it is notable that linear trends, with good correlation coefficients, are always obtained. Moreover, it is notable that both T_{CC} and Tm increase considerably with the increment of the percentage of ttt triads. For example, T_{CC} varies from 79 to 149°C and Tm from 122 to 165°C when the percentage of the ttt triads changes from about 37 to 100%. Therefore, a high percentage of three trans sequences significantly improves the crystalline perfection and crystallizability of PBCHD.

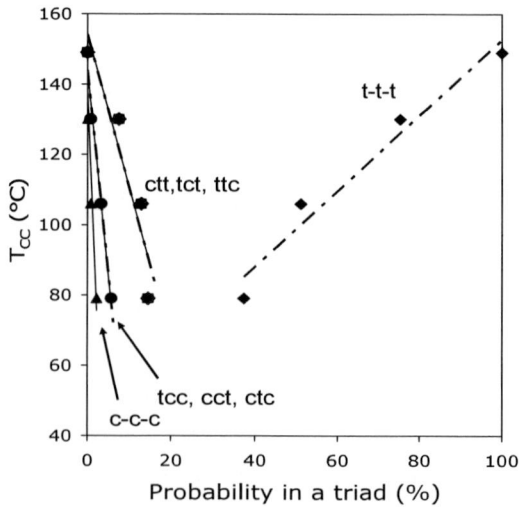

Figure 8. Trends of T_{CC} vs. the probabilities of finding the ttt, ctt, tct, ttc, cct, tcc, ctc, and ccc triads in the sequence of three 1,4-cyclohexylene rings along the PBCHD chain.

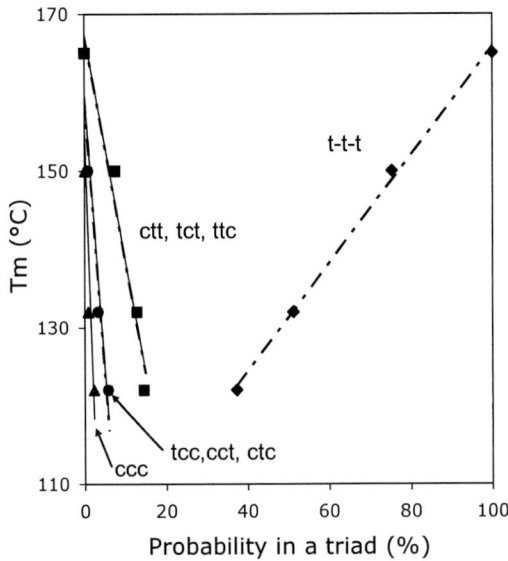

Figure 9. Trends of Tm vs. the probabilities of finding the ttt, ctt, tct, ttc, cct, tcc, ctc, and ccc triads in the sequence of three 1,4-cyclohexylene rings along the PBCHD chain.

Table 4. Probability of finding the ttt, ctt, tct, ttc, cct, tcc, ctc, and ccc triads in a sequence of three cycloaliphatic rings along the PBCHD chain for the semicrystalline samples

Sample	P_{ttt} %	P_{ctt} %	P_{tct} %	P_{ttc} %	P_{cct} %	P_{tcc} %	P_{ctc} %	P_{ccc} %
PBCHD-70	37.3	14.5	14.5	14.5	5.6	5.6	5.6	2.2
PBCHD-80	51.2	12.8	12.8	12.8	3.2	3.2	3.2	0.8
PBCHD-90	75.4	7.5	7.5	7.5	0.7	0.7	0.7	0.1
PBCHD-100	100	0.0	0.0	0.0	0.0	0.0	0.0	0.0

On the other hand, all the other triads (ctt, tct, ttc, cct, tcc, ctc, ccc) cause a sharp and significant decrement of both TCC and Tm: the decrement is even greater when two or, particularly, three rings in cis configuration are present. This result is further evidence that the cis configuration acts as a disturbance in the crystallization process of PBCHD: the presence of "kinks" along the macromolecular chains prevents a high crystalline perfection and high melting peak. The "disturbance" of the cis form is so great that the presence of two rings in trans configuration in the ctt, tct, and ttc triads are not sufficient to favor the formation of stable crystals.

3.1.4.2. PCCD

For PCCD the description of the molecular structure is more complicated than for PBCHD: the two units characteristic of the PCCD are the 1,4-cyclohexanedicarbonyl unit (called E), deriving from DMCD, and the 1,4-cyclohexanedimethyleneoxy unit (called D), deriving from CHDM. Attention has been focused on a sequence of three 1,4-cyclohexylene rings: D-E-D and E-D-E, as represented in Scheme 4.

Firstly, the D-E-D sequence was analysed. Table 5 shows the probabilities of finding the eight possible triads within the chains in the analyzed samples, following eq. (1).

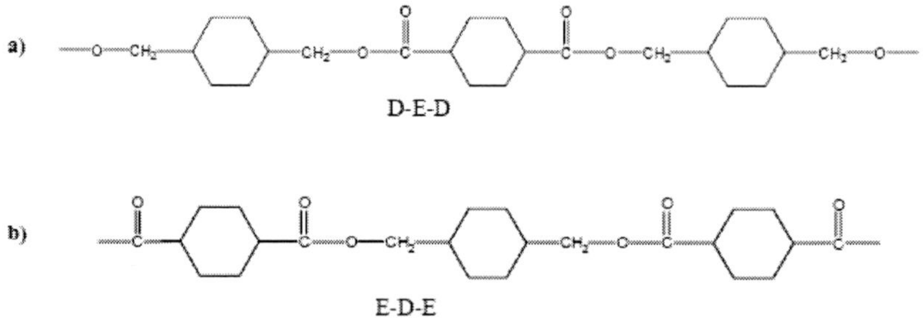

Scheme 4. Representation for PCCD of the different sequences of three 1,4-cyclohexylene rings considered for the determination of the probability of finding the triads reported in Tables 5 and 6.

A first correlation between molecular structure and thermal properties is described in Figure 10, where the Tm values (Table 2) are reported as a function of the probability of finding the ttt, ctt, tct, ttc, cct, tcc, ctc, and ccc triads (Table 5). Also in this case, for the samples presenting a double melting peak, only the first peak is reported, due to the fact that, for the possible melting-recrystallization process, it corresponds to the melting of the crystals actually formed at T_{CC} during the cooling scan.

Firstly, it is noteworthy that linear trends, with good correlation coefficients, are obtained. Moreover, it is notable that Tm increases with the increment of the percentage of four triads (ttt, ctt, ttc, and ctc). For example, Tm varies from 181 to 230°C when the percentage of the ttt triads changes from about 29 to 42%. The same trend is observed for the ctt and ttc triads, i.e.

for triads containing two consecutive rings in trans configuration. This result suggests that a high percentage of two or three trans sequences has a significant role in improving the crystalline perfection of PCCD. The observation is in perfect agreement with the previous discussion and with the literature results.

Table 5. For PCCD probability of finding the ttt, ctt, tct, ttc, cct, tcc, ctc, and ccc triads in the D-E-D sequence

Sample	P_{ttt} %	P_{ctt} %	P_{tct} %	P_{ttc} %	P_{cct} %	P_{tcc} %	P_{ctc} %	P_{ccc} %
D66-E20	10.5	5.4	33.1	5.4	17.1	17.1	2.8	8.8
D66-E50	22.7	11.7	20.9	11.7	10.8	10.8	6.0	5.5
D66-E70	28.7	14.8	14.8	14.8	7.6	7.6	7.6	3.9
D66-E80	35.3	18.2	8.3	18.2	4.3	4.3	9.4	2.2
D66-E90	39.2	20.2	4.4	20.2	2.2	2.2	10.4	1.2
D66-E100	42.3	21.8	1.3	21.8	0.7	0.7	11.2	0.3

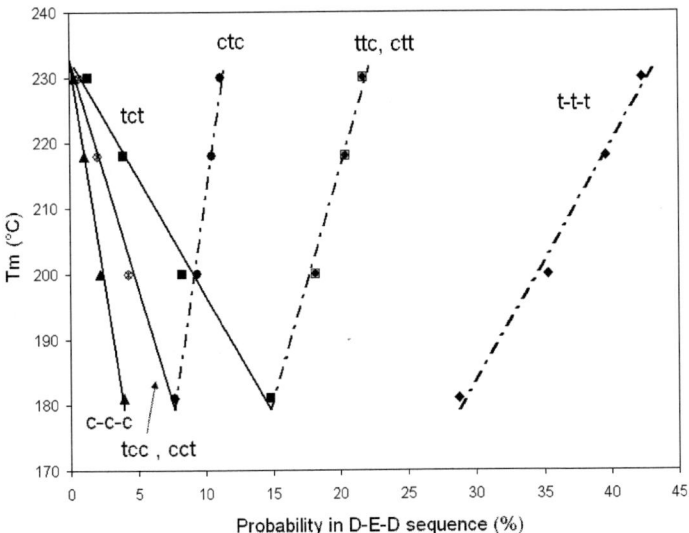

Figure 10. Trends of Tm vs. the probabilities of finding the ttt, ctt, tct, ttc, cct, tcc, ctc, and ccc triads in the D-E-D sequence of PCCD.

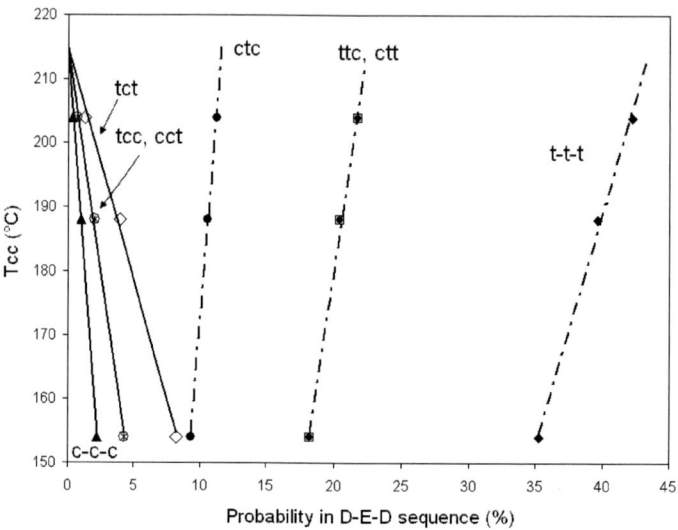

Figure 11. Trends of TCC vs. the probabilities of finding the ttt, ctt, tct, ttc, cct, tcc, ctc, and ccc triads in the D-E-D sequence of PCCD.

Moreover, by observing Figure 10, it may also be seen that the ctc triads, i.e. the triads containing only the unit E in trans configuration, improve Tm significantly, independently of the fact that the other two rings are in cis configuration. On the other hand, a small increment of the number of the the tct triads, where two D units are in trans configuration, causes a rapid decrement of Tm. This result is important because it is a first indication of a different effect of the stereochemistry of the D and E units on the PCCD thermal properties. Indeed, it is sufficient that the E unit is in cis configuration that crystal perfection decreases. Therefore, it seems that the configuration of the E unit has a very important effect in influencing the trend of Tm.

Analysis of the remaining curves shows that Tm decreases as the percentages of the ccc, tcc, and cct triads increase. Also in this case, for the sequences of two or three rings in cis configuration, the result is expected: the presence of "kinks" along the macromolecular chains prevents a high crystal perfection and high melting peak.

Given these results, the effect of the percentage of triads on the crystallization temperature T_{CC}, measured in DSC during the cooling scan from the melt (Table 2), is also significant. In Figure 11 it is evident that the presence of the ttt, ctt, ttc, and ctc triads causes an increment of T_{CC}. The trends are the same with respect to those observed in Figure 10, indicating that a high crystallization capacity and crystal perfection are improved by the presence of two and three sequences of rings in trans and, more significantly, by the presence of an isolated ring, belonging to E unit, in trans configuration. On the contrary, two and three sequences of rings in cis and, more significantly, an isolated ring belonging to E unit, in cis configuration inhibit the crystallization process.

Therefore, in order to emphasise the different effect of the stereochemistry of the C_6 ring, depending on whether it belongs to E or D unit, we have analyzed the molecular structure of the PCCD samples focusing on the E -D-E sequences (Scheme 4, b). In this case two E units and only one D unit have been considered. The probability of finding the eight possible triads (ttt, ctt, tct, ttc, cct, tcc, ctc, ccc) has been calculated and the results reported in Table 6.

Table 6. Probability of finding the ttt, ctt, tct, ttc, cct, tcc, ctc, and ccc triads in the E-D-E sequence

Sample	P_{ttt} %	P_{ctt} %	P_{tct} %	P_{ttc} %	P_{cct} %	P_{tcc} %	P_{ctc} %	P_{ccc} %
D66-E20	3.8	12.0	2.0	12.0	6.2	6.2	38.1	19.6
D66-E50	17.8	16.5	9.1	16.5	8.4	8.4	15.2	7.8
D66-E70	28.7	14.8	14.8	14.8	7.6	7.6	7.6	3.9
D66-E80	43.3	10.2	22.3	10.2	5.2	5.2	2.4	1.2
D66-E90	53.5	5.9	27.5	5.9	3.1	3.1	0.7	0.3
D66-E100	62.1	1.9	32.0	1.9	1.0	1.0	0.1	0.03

Figure 12 describes the trend of Tm as a function of the probability of finding the different triads in the E-D-E system. Now an important difference may be seen with respect to the trends reported in Figure 10 and 11: the increment of Tm occurs only for two kinds of sequences, i.e. ttt and tct. All the other sequences (ccc, cct, ctc, tcc, ttc, and ctt) induce a decrement of Tm. It is notable that for T_{CC} the trend is exactly the same.

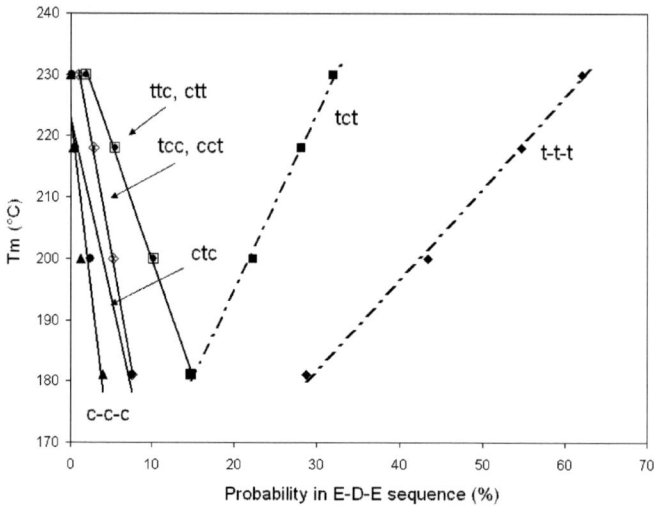

Figure 12. Trends of Tm vs. the probabilities of finding the ttt, ctt, tct, ttc, cct, tcc, ctc, and ccc triads in the E-D-E sequence of PCCD.

Therefore, the ccc, cct, ctc, tcc, ttc, and ctt triads are responsible of lower crystal perfection and lower tendency to crystallise. Then, in this case, the presence of two consecutive rings in trans configuration (triads ttc and ctt) is not determinant in improving crystallizability. Indeed, the configuration of the ring of E unit appears to be the most important factor influencing crystallisation capacity: its trans configuration favours crystallization, whereas its cis configuration is unfavourable.

As regards the glass transition, Figure 13 describes the trend of Tg as a function of the probability of finding the different triads in the D-E-D sequences. It is notable that the same results discussed for Tm and T_{CC} (Figures 10 and 11) are obtained. Analogously, for the E-D-E sequences the trend of Tg vs the probability of finding the different triads is exactly the same of that shown in Figure 12 for Tm and that obtained for T_{CC}. As a conclusion, Tg increases as the % of ttt triads increases and when the 1,4-cyclohexylene ring in E unit is in trans configuration.

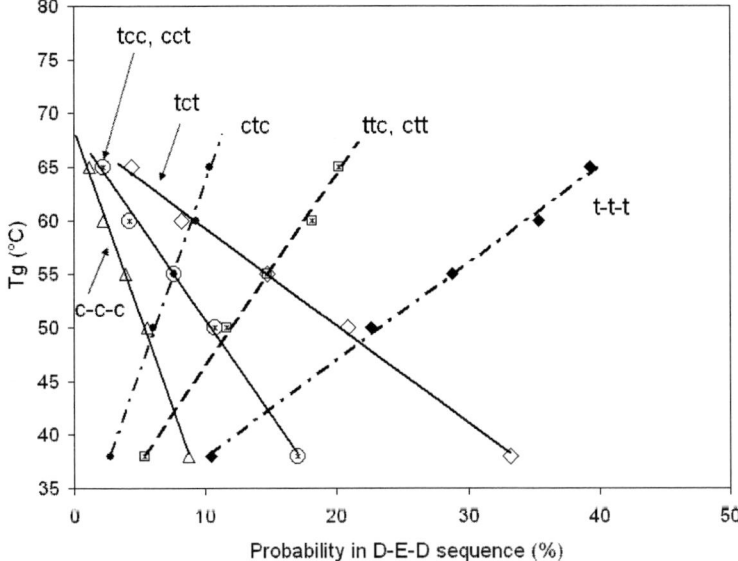

Figure 13. Trends of Tg vs. the probabilities of finding the ttt, ctt, tct, ttc, cct, tcc, ctc, and ccc triads in the D-E-D sequence.

The results obtained can be discussed in terms of chain mobility. It is noteworthy that in unit E the 1,4-cyclohexylene moiety is linked to two ester groups, where the carbon atoms are characterized by sp^2 hybridisation, with planar geometry and bond angles of 120°. This structure has a certain degree of rigidity. On the other hand, in unit D the rings are connected to two $-CH_2-$ groups, which are characterised by carbon atoms with sp^3 hybridisation, tetrahedral geometry and bond angles of about 109°. In this case, each carbon atom possesses a higher mobility and can assume different but equivalent orientations in space. Therefore, the most rigid structure, the E unit, is found to have the highest influence in determining the thermal properties.

Moreover, the configurations of the two sequences D-E-D are shown in Scheme 5: ctc and tct. The influence of these two sequences on the thermal transitions, previously discussed, is significant because the first sequence (a) causes an increment of Tg, T_{CC} and Tm, whereas the second (b) induces a decrement of the Tg, T_{CC} and Tm. This means that when the aliphatic ring in E unit is in trans configuration an improvement of the thermal properties occurs, whereas when the C_6 ring is in cis configuration the thermal properties are lower.

In particular, the rigidity of E unit in trans configuration (a) gives a high symmetry to ctc triad, improving chain packing, crystalline order and also Tg. On the other hand, the cis configuration of E unit (b) is characterized by a high steric hindrance of the two substituent groups. This leads to less probability of chain folding and crystallizability for tct than ctc triad.

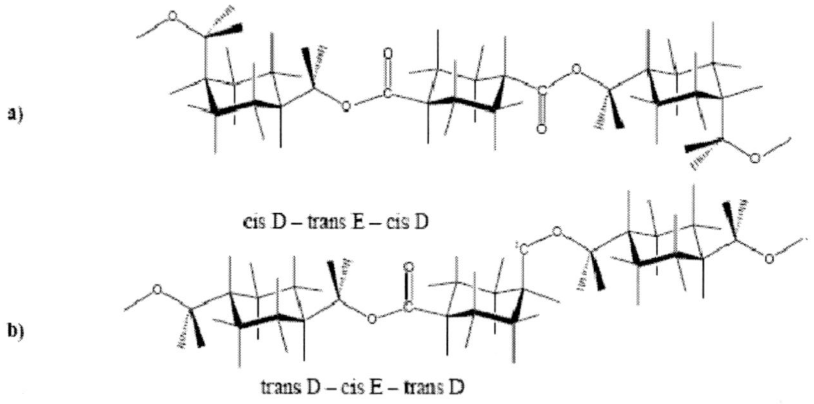

a)

cis D – trans E – cis D

b)

trans D – cis E – trans D

Scheme 5. Two configurations of D-E-D sequence.

3.2. Preparation and Characterization of Copolymers

3.2.1. Molecular Characterization

Some novel aliphatic, random copolyesters were synthesized from DMCD, BD, and DMA by following the procedure described in the Experimental Part. They are called (4-6)-co-PBCHDxx-a/b, where xx indicates the percentage of the aliphatic rings, derived from DMCD, in trans configuration and a/b is the feed molar ratio of the DMA/DMCD.

The molecular structure is reported in Scheme 6, while Table 7 describes all the samples prepared.

The chemical structure was analyzed by ^{1}H NMR spectroscopy. For all the samples the spectra were found to be consistent with the expected structures. The composition was calculated, but it was not possible to evaluate the degree of randomness and the average length of the sequences. However, since we observed no differences in the reactivity of BD with respect to the two diesters during the synthesis of the two homopolymers (4-6 and PBCHD), we hypothesize that the copolyesters have a random distribution of the sequences along the chain, as demonstrated for aliphatic-aromatic (4-12)-co-PBT samples [39].

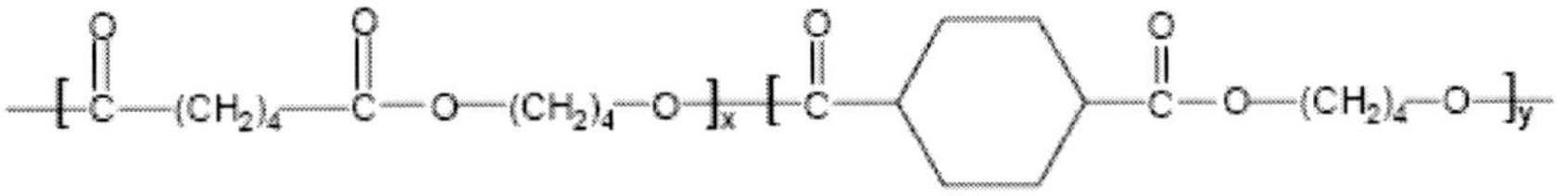

(4-6)-co-PBCHD

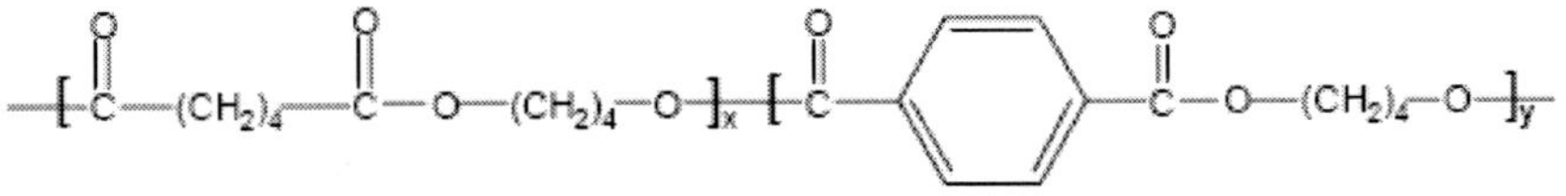

PBTA – trade name: Ecoflex

Scheme 6. Molecular structures of the novel (4-6)-co-PBCHD copolyesters and Ecoflex.

Table 7. Molecular characteristics of the copolyesters

Sample	(4-y)/PBCHD molar ratio[a]	trans % of the ring in the polymer[a]	$Mw \cdot 10^{-3}$ [b]	Mw/Mn [b]
4-6	-	-	90.0	2.5
Ecoflex	-	-	121.0	2.6
(4-6)-co-PBCHD$_{100}$-70/30	64/36	100	95.4	2.1
(4-6)-co-PBCHD$_{100}$-50/50	47/53	100	73.6	2.3
(4-6)-co-PBCHD$_{100}$-30/70	24/76	100	97.4	2.2
(4-6)-co-PBCHD$_{90}$-70/30	65/35	88	84.9	2.0
(4-6)-co-PBCHD$_{90}$-50/50	47/53	87	87.8	2.2
(4-6)-co-PBCHD$_{90}$-30/70	24/76	90	99.4	2.4
(4-6)-co-PBCHD$_{70}$-70/30	65/35	70	86.3	2.0
(4-6)-co-PBCHD$_{70}$-50/50	47/53	68	88.5	2.3
(4-6)-co-PBCHD$_{70}$-30/70	24/76	72	95.3	2.4
(4-6)-co-PBCHD$_{50}$-70/30	65/35	53	76.4	2.1
(4-6)-co-PBCHD$_{50}$-50/50	46/54	56	65.1	2.1
(4-6)-co-PBCHD$_{50}$-30/70	24/76	56	128.3	2.2

[a] Calculated by ^{1}H NMR.
[b] Measured by GPC in $CHCl_3$.

From the ratio of the contributions of the protons of the cycloaliphatic ring in trans and in cis configurations, the trans percentage of the cycloaliphatic units inside the copolymers has been also determined.

As a result, the copolyesters differentiate for molar composition and for trans mol % of the PBCHD units (see Table 7). In particular, the trans content varies from 50 to 100 mol %, as in the case of the corresponding PBCHD homopolymers previously described. It is also evident that the molecular weights of all the samples are very high. In this way, the samples can conveniently be compared in order to study their thermal behavior.

3.2.2. Analysis of the Phase Behaviour of Copolymers

One of reasons of the great interest towards (4-6)-co-PBCHD copolyesters is due to the fact that they are the corresponding aliphatic materials of the poly(butylene terephthalate-co-butylene adipate) (PBTA) (see Scheme 6). This is an aliphatic-aromatic copolymer, commercialized with the trade name of Ecoflex. PBTA is characterized by the presence of terephthalate groups and, thus, for the reasons already discussed, it has relatively high transition temperatures: Tg is located at about -30°C and melting temperature at 120°C. Therefore, the introduction of aromatic rings to the aliphatic chains of 4-6 induces a notably improvement in thermal characteristics. Moreover, PBTA has good mechanical properties and excellent thermal stability [40]. As regards its biodegradation, Witt et al. concluded that there is no indication of environmental risk when it is involved in the composting process [41], in spite of the presence of not-biodegradable units. Therefore, Ecoflex is described as biodegradable plastic ideal for trash bags and disposable packaging.

Indeed, the novel (4-6)-co-PBCHD copolyesters are characterized by mechanical properties not too different with respect to those of Ecoflex, as shown in Figure 14, and, also the thermal properties are not poor, as discussed later (see Table 8). Moreover, since our copolymers are all-aliphatic, their biodegradability is potentially really good [12]. As a consequence, the new (4-6)-co-PBCHD copolyesters are excellent candidates to cover important roles in the future applications of polymers.

As to the relationships between stereochemistry and phase behavior, Figure 15 and 16 reports some examples of DSC curves (cooling and heating scans), obtained on (4-6)-co-PBCHD-30/70 samples, with trans percentage varying from 100 to 50 mol %.

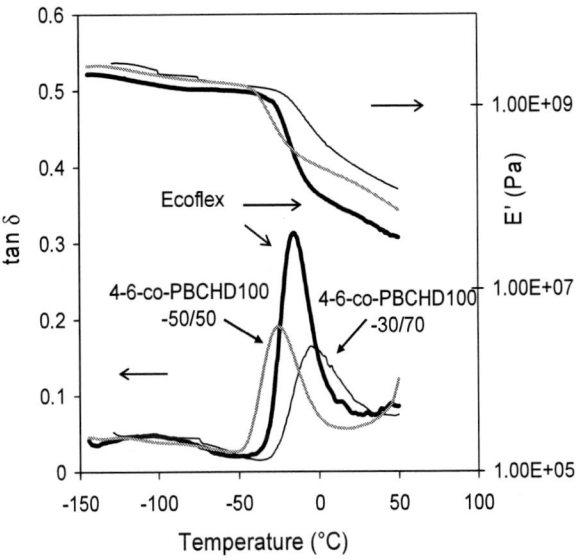

Figure 14. DMTA spectra of Ecoflex, (4-6)-co-PBCHD$_{100}$-70/30, and (4-6)-co-PBCHD$_{100}$-50/50 copolymers.

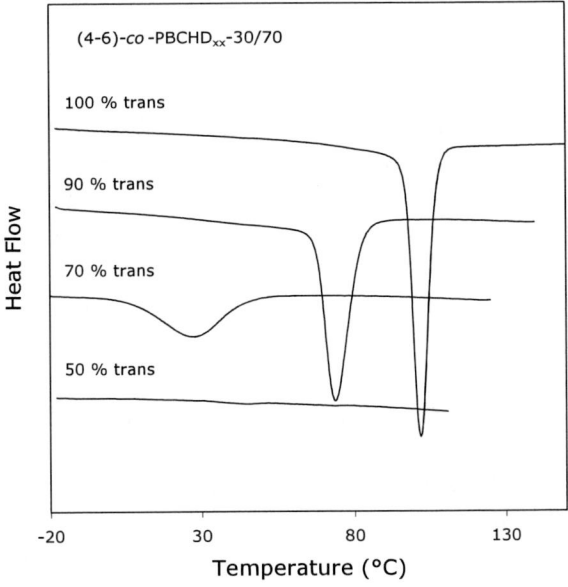

Figure 15. DSC traces of the cooling scans at 10° C·min^{-1} for the 4-6-co-PBCHD copolymers.

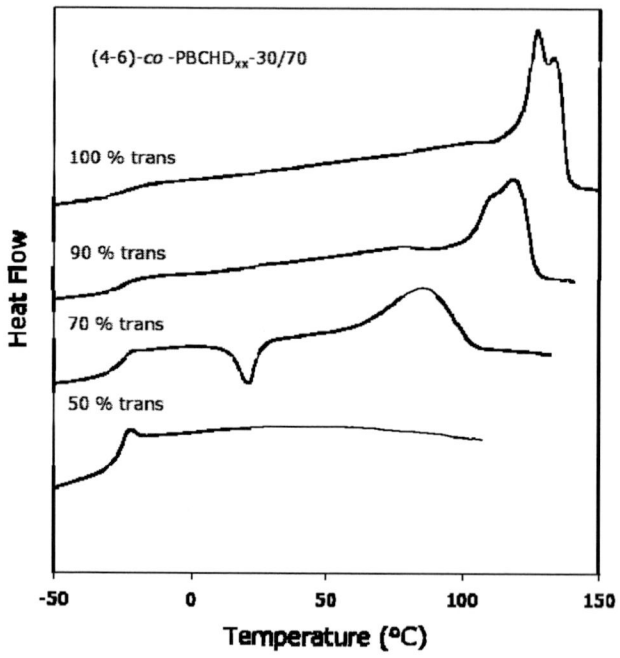

Figure 16. DSC traces of the 2nd heating scans at 10°C·min^{-1} for the 4-6-co-PBCHD copolymers.

In this case, the copolymers at high percentage of PBCHD units are taken in account, because the effect of the stereochemistry of the cycloaliphatic ring is more evident.

As already discussed for PBCHD homopolymers, the capacity of the samples to crystallize is found to be to strongly modified. For example, the copolymer containing PBCHD units with 50 mol % of trans isomer is a fully amorphous material, which is not able to crystallize either during the cooling scan from the melt or during the subsequent heating scan. By increasing the trans content to 70 mol %, the copolymer gains the capacity to organize itself into a more ordered structure, partially during the cooling scan and partially during the heating scan. Samples at high trans contents (90 and 100 mol %) crystallize with narrow peaks, reaching level of crystallinity similar to that of the homopolymer, although TCC values are significantly lower (T_{CC} = 97°C for (4-6)-co-PBCHD$_{100}$-30/70 and 149°C for PBCHD$_{100}$). Correspondently, the Tm values also increase with the trans content, as evident from an analysis of all the data of Table 8.

Different observations can be made for the Tg values, reported in Table 1 and 8. Indeed, if the Tg data of PBCHD samples are compared as a function of the trans content, it is evident that Tg decreases with the decrement of the trans content, due to the low flexibility of the 1,4-cyclohexylene ring in trans configuration. From $PBCHD_{100}$ to $PBCHD_{50}$ Tg varies from 10 to -7°C. For the (4-6)-co-PBCHD copolymers, instead, this trend is not observed. For example, all the (4-6)-co-PBCHD-70/30 samples have experimental Tg values of about -50°C, independently of the fact that the trans content varies from 100 to 50 mol %.

Table 8. Thermal data of the copolyesters

Sample	T_{CC} [a] °C	ΔH_{CC} [a] J·g^{-1}	T_g [b] °C	T_{ch} [b] °C	ΔH_{ch} [b] J·g^{-1}	T_m [b] °C	ΔH_m [b] J·g^{-1}
4-6	32	67	-58	-	-	52 and 57	70
Ecoflex	80	21	-31	-	-	124	11
(4-6)-co-PBCHD$_{100}$-70/30	14	28	-49	-	-	53	24
(4-6)-co-PBCHD$_{100}$-50/50	66	33	-43	-	-	94	32
(4-6)-co-PBCHD$_{100}$-30/70	97	41	-23	-	-	125 and 134	40
(4-6)-co-PBCHD$_{90}$-70/30	2	30	-49	-	-	43	27
(4-6)-co-PBCHD$_{90}$-50/50	41	25	-40	-	-	80	26
(4-6)-co-PBCHD$_{90}$-30/70	74	29	-25	-	-	119	27
(4-6)-co-PBCHD$_{70}$-70/30	-20	4	-50	-9	19	30	23
(4-6)-co-PBCHD$_{70}$-50/50	-	-	-40	19	14	56	16
(4-6)-co-PBCHD$_{70}$-30/70	27	16	-25	21	5	86	22
(4-6)-co-PBCHD$_{50}$-70/30	-	-	-52	-2	5	18	6
(4-6)-co-PBCHD$_{50}$-50/50	-	-	-41	-	-	-	-
(4-6)-co-PBCHD$_{50}$-30/70	-	-	-26	-	-	-	-

[a] Measured by DSC (cooling scan at 10° C·min^{-1}).
[b] Measured by DSC (2nd heating scan at 10° C·min^{-1}).

All the (4-6)-co-PBCHD-50/50 copolymers have Tg values of about -40°C and all the (4-6)-co-PBCHD-30/70 copolymers have Tg values of about -25°C. In this case, the effect of the stereoregularity of the aliphatic ring on the chain flexibility seems to be absent. Indeed, it is reasonable that an effect of the cis/trans ratio of the 1,4-cyclohexylene ring should be present mainly in the copolymers rich in PBCHD, i.e. in 30/70 copolymers. However, the presence of the PBCHD crystalline phase [21] creates a more rigid matrix and originates an amorphous phase whose composition is not perfectly correspondent to the theoretical one, but richer in 4-6 units. The 4-6 units can lead to a greater flexibility of the chain and, thus, to lower Tg data. Therefore, opposite effects are present: the trans isomer and the crystalline phase increase the rigidity of the system and cause an increment of Tg, whereas the 4-6 units, which are preferentially in the amorphous state, induces a decrement of Tg. As a result, the final Tg value could be a constant.

CONCLUSION

The thermal behaviour of polyesters and copolyesters, containing the 1,4-cyclohexylene units, have been analyzed in order to highlight the correlations existing between stereochemistry of the aliphatic ring and phase behaviour. It is noteworthy that significant relationships have been found for all the materials here analyzed (PBCHD, PCCD, (4-6)-co-PBCHD).

In particular, with the increment of the trans content the polymers change from completely amorphous to semicrystalline material. Correspondingly, Tg increases of about 20-25°C in PBCHD and PCCD with the increment of the trans content from 20 to 100 mol%. At the same time, Tm increases of about 40-50°C with the increment of the trans amount from 70 to 100 mol%. Moreover, the polymers in all trans conformation are able to reach a very high level of crystallinity and crystal perfection. This behaviour has been attributed to the better chain packing which occurs in the presence of the more "stretched" trans conformation. Indeed, the rings in cis conformation cause "kinks" and a lower degree of symmetry.

An analysis of the molecular structure of PCCDs shows that the stereochemistry of 1,4-cyclohexylene ring deriving from DMCD is the main element responsible for the thermal properties. This result is due to the higher rigidity of the 1,4-cyclohexanedicarbonyl unit with respect to 1,4-cyclohexanedimethyleneoxy unit, deriving from CHDM. Therefore, when the most rigid structure is in trans conformation, Tg, Tm and crystallization rate tend to increase, whereas the cis conformation of the 1,4-cyclohexanedicarbonyl unit tends to induce a decrement of Tg and crystallizability.

Significant results have been found in copolymers too: in this case, it is possible to easily modulate the final properties of the materials not only by changing the molar composition, but also varying the cis/trans ratio of the alicyclic units inside the chains.

As a conclusion, PBCHD, PCCD, and (4-6)-co-PBCHD copolymers present the great advantages of having cycloaliphatic units which, according to their stereochemistry, allow the material to show modulated thermal properties and good physical performance, and a fully aliphatic main chain, which could provide the favourable biodegradability and resistance to weather, light, heat and water, typical of the aliphatic polyesters.

REFERENCES

[1] El-Hadi, A.; Schnabel, R.; Straube, E.; Müller, G.; Henning, S. *Polymer Testing* 2002, 21, 665-674.
[2] Berti, C.; Celli, A.; Marchese, P.; Marianucci, E.; Barbiroli, G.; Di Credico, F. *e-Polymers* 2007, 057, 1-17.
[3] Liu, Y.; Turner, R. J. *Polym. Sci., Part A: Polym. Chem.* 2010, 48, 2162-2169.
[4] Jackson Jr., W. J.; Darnell, W. R. *US Patent* 4327206, 1982.
[5] Jackson Jr., W.J.; Darnell, W. R. *US Patent* 4342862, 1982.
[6] Brunelle, D. J.; Jang, T. *U.S. Patent* 6, 084, 055, 2000.
[7] Brunelle, D. J.; Berti, C.; Celli, A.; Colonna, M.; Fiorini, M.; Marianucci, E.; Messori, M.; Pilati, F.; Sisti, L.; Marchese, P. *U.S. Patent* 6, 828, 410, 2003.
[8] Brunelle, D. J.; Jang, T. *Polymer* 2006, 47, 4094-4104.
[9] Turner, S. R.; Seymour, R. W.; Dombroski J. R.. In: Scheirs, J.; Long, T. E. Editors. *Modern polyesters*. Chichester, Wiley; 2003, 267-292.
[10] Berti, C.; Binassi, E.; Colonna, M.; Fiorini, M.; Kannan, G.; Kanaram, S.; Mazzacurati, M. US 20100168461.
[11] Bechthold, I.; Bretz, K.; Kabasci, S.; Kopitzky, R.; Springer, A. *Chem. Eng. Technol.* 2008, 31, 647-654.
[12] Berti, C.; Celli, A.; Marchese, P.; Barbiroli, G.; Di Credico, F.; Verney, V.; Commereuc, S. *Europ. Polym. J.* 2009, 45, 2402-2412.

[13] Kricheldorf, H. R.; Schwarz, G. *Makromol. Chem.* 1987,188,1281-1294.
[14] Vanhaecht, B.; Teerenstra, M. N.; Suwier, D. R.; Willem, R.; Biesemans, M.; Koning, C. E. *J. Polym. Sci. Part A: Polym. Chem.* 2001, 39, 833-840.
[15] Vanhaecht, B.; Rimez, B.; Willem, R.; Biesemans, M.; Koning, C. E. *J. Polym. Sci. Part A: Polym. Chem.* 2002, 40, 1962-1971.
[16] Koning, C.; Vanhaecht, B.; Willem, R.; Biesemans, M.; Goderis, B.; Rimez, B. *Macromol. Symp.* 2003, 199, 431-442.
[17] Vanhaecht, B.; Willem, R.; Biesemans, M.; Goderis, B.; Basiura, M.; Magusin, P.C.M.M.; Dolbnya, I.; Koning, C. E. *Macromolecules* 2004, 37, 421-428.
[18] Joseph, M. D.; Savina, M. R.; Harris, R. F. *J. Appl. Polym. Sci.* 1982, 44, 1125-1134.
[19] Berti, C.; Celli, A.; Marchese, P.; Marianucci, E.; Barbiroli, G.; Di Credico, F. *Macromol. Chem. Phys.* 2008, 209, 1333-1344.
[20] Berti, C.; Binassi, E.; Celli, A.; Colonna, M.; Fiorini, M.; Marchese, P.; Marianucci, E.; Gazzano, M.; Di Credico, F.; Brunelle, D. J. *J. Polym. Sci. Part B: Polym. Phys.* 2008, 46, 619-630.
[21] Berti, C.; Celli, A.; Marchese, P.; Marianucci, E.; Sullalti, S.; Barbiroli, G. *Macromol. Chem. Phys.* 2010, 211, 1559-1571.
[22] Sánchez-Arrieta, N.; Martínez di Ilarduya, A.; Alla, A.; Muñoz-Guerra, S. *Europ. Polym. J.* 2005, 41,1493-1501.
[23] Wang, L.; Xie, Z.; Bi, X.; Wang, X.; Zhang, A.; Chen, Z., Zhou, J.; Feng, Z. *Polym. Degrad. Stab.* 2006, 91, 2220-2228.
[24] Sandhya, T. E.; Ramesh, C.; Sivaram, S. *Macromolecules* 2007, 40, 6906-6915.
[25] Berti, C.; Bonora, V.; Colonna, M.; Lotti, N.; Sisti, L. *Europ. Polym. J.* 2003, 39,1595-1601.
[26] Colonna, M.; Berti, C.; Binassi, E.; Celli, A.; Fiorini, M.; Marchese, P.; Messori, M.; Brunelle, D. J. submitted to *Polym. Int.*, 2010.
[27] Matsuda, H.; Nagasaka, B.; Asakura, T. *Polymer* 2003, 44, 4681-4687.
[28] Marchese, P.; Celli, A.; Fiorini, M. *J. Polym. Sci. Part B: Polym. Phys.* 2004, 42, 2821-2832.
[29] Tonelli, A. E. *J. Polym. Sci., Polym. Lett. Ed.* 1973, 11, 441-447.
[30] Tonelli, A. E. *J. Polym. Sci., Polym. Phys.* 2002, 40, 1254-1260.
[31] Gonzalez, C. C.; Riande, E.; Bello, A.; Perefia, J. M. *Macromolecules* 1988, 21, 3230-3234.
[32] Ki, H.C.; Ok Park, O. *Polymer* 2001, 42,1849-1861.
[33] Kwolek, S. L. ; Luise, R. R. *Macromolecules* 1986, 19, 1789-1796.

[34] Osman, M. A. *Macromolecules 1986*, 19, 1824-1827.
[35] Reck B.; Ringsdorf, H.; Gardner, K.; Starkweather Jr., H. *Makromol. Chem.* 1989, 190, 2511-2526.
[36] Lenz, R.W.; Go S. *J. Polym. Sci., Polym. Chem. Ed.* 1973, 11, 2927-2946.
[37] Srinivasan, R.; McGrath, J. E. *Polym. Prepr.* 1992, 33, 503-504.
[38] Ridgway, J. S. *J. Polym. Sci., Polym. Chem.* 1970, 8, 3089-3111.
[39] Berti, C.; Celli, A.; Marchese, P.; Barbiroli, G.; Di Credico, F.; Verney, V.; Commereuc, S. *Europ. Polym. J.* 2008, 44, 3650-3661.
[40] Herrera, R.; Franco, L.; Rodríguez-Galán, A.; Puiggalí, J. *J. Polym. Sci. Part A: Polym. Chem.* 2002, 40, 4141-4157.
[41] Witt, U.; Einig, T.; Yamamoto, M.; Kleeberg, I.; Deckwer, W. -D.; Müller, R.-J. *Chemosphere* 2001, 44, 289-299.

In: New Developments in Polymers Research ISBN: 978-1-61942-915-4
Editors: E. O. Bradley and M. I. Lane © 2012 Nova Science Publishers, Inc.

Chapter 3

THERMO- AND PH-SENSITIVITY OF POLY (*N*-VINYLPYRROLIDONE) IN WATER MEDIA

N. I. Pakuro[*], *A. A. Arest-Yakubovich, B. I. Nakhmanovich and F. Kh. Chibirova*

Karpov Institute of Physical Chemistry
per. Obukha 3-1/12, str. 6, Moscow, Russia

ABSTRACT

In recent years, a number of polymers that undergo phase separation in water solutions on temperature rising are studied. These polymers are characterized by lower critical solution temperatures (LCST). Poly(*N*-vinylpyrrolidone) (PVP) is not thermo- or pH- sensitive under usual conditions. However, since this polymer is widely used, especially in medicine, several studies are dedicated to the problem of making this polymer stimuli-responsive, too. In the review, the phase behavior of PVP in water solutions under various conditions is covered. The phase behavior of PVP-containing copolymers and hydrogels are described. The effect of the addition of salts, including transition metal ones, on the PVP phase separation temperature is considered, the attention being paid to the different influence of anions and cations on this value. It is known that PVP readily forms complexes with many organic and inorganic compounds. Examples of such complex formation effects on cloud points of the polymer solutions are given. The phase behavior of PVP is

---

[*] E-mail address: pakuro@cc.nifhi.ac.ru

compared with that of poly(*N*-vinylcaprolactam), a PVP close analog, which is a well-known thermosensitive polymer.

INTRODUCTION

Stimuli-responsive water-soluble polymers are the subject of wide research because of their possible applications in medicine (drug release), biotechnology, and many other fields. Temperature responsive polymers undergo phase separation on increasing the solution temperature, i.e., they have a lower critical solution temperature (LCST). Poly(*N*-isopropylacrylamide) (PNIPAM), other polyacrylamides, poly(*N*-vinylcaprolactam) (PVCL) are typical examples of such polymers. A number of reviews on the subject are published [1]-[5]. Some thermosensitive polymers, for example, poly(*N*-vinylcaprolactam) - co methacrylic acid, are also pH responsive [6].

Poly(*N*-vinylpyrrolidone) (PVP) is the closest analog of PVCL, but while the latter phase separates at temperatures of 32-39 °C, PVP exhibits thermosensitivity only under pressure higher than 1 kbar [7]. It is interesting that in the pressure range 2-4 kbar, both LCST, and upper critical solution temperature (UCST) are observed. PVP is widely used in medicine and pharmacology due to its biocompatibility and very low toxicity. Nowadays, it finds a new application as a stabilizer of metal and metaloxide sols in the sol-gel nanotechnology [8]-[10]. In this connection, tries have been taken at making PVP stimuli-responsive, too.

THERMOSENSITIVITY OF PVP SOLUTIONS, CONTAINING SALTS

The most usual way of controlling the temperature of phase separation $T_{ph.s}$ of thermosensitive polymers is addition salts into their solutions. Influence of inorganic salts on properties of PVCL water solutions has been studied by Kirsh [3], [4], [11]. Addition of various salts was shown to result both in an increase and decrease in polymer $T_{ph.s}$, depending on the nature and concentration (*c*) of the salt. Sulfates, carbonates, and phosphates decrease $T_{ph.s}$ of PVCL most effectively. This salting out effect can be explained by weakening hydrogen bonds in the polymer - hydrate complex that increases hydrophobic interactions in the polymer chain and enhances coil-to-globule transition. Some authors believe that this phenomenon is caused by a change in the water structure upon salt addition [3], [4].

The same effects have been observed for water solutions of PVP. $T_{ph.s}$ of this polymer ($M = 4 \times 10^4$) was found to linearly decrease with increasing alkali metal salts concentrations in 1% polymer solutions [12]. The influence of anions corresponds to the following row: F^- (1-2) < $H_2PO_4^-$ (0.8-1.4) < CO_3^{2-} (0.4-0.7) < SO_4^{2-} (0.17-0.27). Concentration ranges (mol/l) where $T_{ph.s}$ falls down from 80 to 25 °C are given in brackets. Addition of NaCl or NH_4NO_3 does not result in phase separation even at their concentrations higher than 3 mol/l.

These data are in agreement with results of viscometric measurements [13] where addition of sulfates, diphosphates, and tetraborates is shown to decrease the intrinsic viscosity of PVP solutions more strongly than NaCl and $NaClO_4$ do. This indicates the greater compactization of macromolecules in the first case that facilitates phase separation.

θ-temperatures for PVP solutions, containing these salts, have been obtained from their cloud point temperatures [14]. Extrapolation of the $T_{ph.s}$ - c plots to zero concentration gives $T_{ph.s}$ value close to 170 °C which is in agreement with that obtained earlier [3].

For PVCL, it is shown that a change in $T_{ph.s}$ of its solutions is determined much more by the nature of anions rather than that of cations. Apparently, this is the case for PVP, too. The effectiveness of salts with the same anion (sulfates) is found to depend on the cation charge not sequentially and increase slowly in the following row: Zn^{2+}(0.63-0.9) < Mg^{2+} (0.55-0.87) < Al^{3+}(0.23-0.37) < Na^+(0.17-0-27) [15]. In brackets, the concentration ranges (mol/l), where $T_{ph.s}$ values of PVP solutions decrease linearly from 80 down to 25 °C, are presented. This effect is known for other stimuli-responsive polymers, but is not theoretically considered.

$T_{ph.s}$ values of PVP solutions are shown to depend on polymer concentration that differs this polymer from PVCL. For the latter, the difference in $T_{ph.s}$ for 1 and 20% solutions is not greater than 1-2 °C while for PVP of the same concentrations in 1.5 M KF solution, $T_{ph.s}$ = 52 and 28 °C, respectively [12].

Effect of transition metal chlorides on the properties of PVP and PVCL water solutions is also studied [16]. Addition of $ZnCl_2$ to PVP solutions is shown to result in polymer phase separation upon heating. In the presence of $CdCl_2$ this phenomenon is absent, but on addition of HCl into PVP solutions, containing $CdCl_2$, phase separation also takes place. A very sharp decrease in

$T_{\text{ph.s}}$ is observed if NaCl (KCl) and a small amount of HCl are introduced into PVP - CdCl$_2$ solution. Addition of CuCl$_2$ to PVP solution is effective in the presence of alkali metal chlorides, the effect being more pronounced if HCl is also added. The same phenomena are observed for PVCL. In the authors opinion, the effects described are connected with complex anions MtCl$_4^{2-}$ which are known to form in the transition metal water solutions, containing HCl and/or alkali metal halogenides [17]. The authors [16] believe that in complex systems formed, which include bivalent anions MtCl$_4^{2-}$, complexes of transition metal cations with PVP, and free ions, screening electrostatic interaction between such chains, bivalent anions are responsible for binding chain units with participation of H$_2$O molecules and cations, that facilitates polymer globulization and phase separation on heating. If HNO$_3$ is used instead of HCl, the effects described above are absent that confirms this suggestion.

PHASE SEPARATION OF PVP SOLUTIONS IN THE PRESENCE OF ORGANIC COMPOUNDS CAPABLE TO FORM COMPLEXES WITH THE POLYMER

Complex formation is one of the ways of polymer hydrophobization that facilitates phase separation in their solutions on heating. PVP forms complexes with many organic compounds, including phenols and other aromatic compounds, a number of alcohols and organic acids [18]. The effects of these compounds on PVP solution properties have been studied in comparison with those of PVCL. Low-molecular-mass organic acids, such as formic, acetic, and oxalic ones, are shown to decrease $T_{\text{ph.s}}$ of PVCL [3], [11], but are ineffective in the case of PVP. However, addition of isovaleric and isobutyric acids, which have longer hydrocarbon chains, results in phase separation in PVP solutions [19], indicating an important role of hydrophobic interactions in these systems. Hydrogen bonding between OH groups of acids and C = O groups of polymer lactam rings also takes place. Apparently, these two types of interaction violate the polymer-hydrate complex that facilitates phase separation.

Even the stronger effect on behavior of PVP water solutions is observed if trichloroacetic acid (TCA), which is also known to form complexes with PVP [18], is used as an additive [19]. In the TCA concentration range 0.17-0.30 mol/l, $T_{\text{ph.s}}$ drops from 80 to 10 °C. This system turned out to be also pH-responsive. HCl introduction into solutions, containing TCA, causes even more drastic decrease in $T_{ph.s}$. The same is shown for PVCL while HCl addition into PVCL solutions in the absence of TCA slightly increases $T_{ph.s}$. These phenomena are supposed to be connected with the fact that non-ionized form of TCA takes part in the complex formation via hydrogen bonds between COOH groups of TCA and C = O groups of PVP (PVCL).

To prove this suggestion, the dependence of $T_{ph.s}$ on the concentrations of non-ionized TCA $c(1-\alpha)$ in PVP-TCA solutions was elucidated. The dissociation degree α was calculated for cases of either the presence or absence of HCl in the solution, using the known value of the TCA dissociation constant K_a. All $T_{ph.s}$ vs. [HCl] plots obtained at various TCA concentrations and $T_{ph.s}$ vs. c plot for the case of HCl absence turned out to merge when replotted on the coordinates $T_{ph.s}$ vs. $c(1-\alpha)$. The same is shown to be true for PVCL. These results corroborate the supposition that complexes between the polymers studied and the non-ionized form of TCA cause phase separation in these systems.

The behavior of PVP in water - trifluoroethanol (TFE) mixtures has been studied [20]. It was found that at TFE concentrations 1.3-1.6 mol/l (solution of alcohol in water), transitions of LCST type are observed, i.e., the polymer precipitates on heating. At TFE concentrations in the range 5-6 mo/l (solution of water in alcohol), UCST transitions take place, i.e., the polymer dissolves on heating. In the intermediate field, the system is shown to consist of two phases at any temperatures. Thus, in this system, the phenomenon of so-called cononsolvency is observed, i.e., the polymer entirely dissolves both in water and TFE taken separately, but does not dissolve in their mixtures. TFE and water are completely mixable in any proportions. The cononsolvensy phenomenon was reported earlier for PNIPAM [21], where it was explained by stronger interaction between water and cononsolvent than that between the polymer and any solvent, resulting in decomposition of polymer-hydrate complex and polymer sedimentation. For PVCL it was not reported.

PHASE SEPARATION IN WATER SOLUTIONS OF PVP COPOLYMERS AND DERIVATIVES

Copolymerization of VP with less hydrophylic monomers is a means to obtain thermoresponsive polymers. Random copolymer of VP and VCL (M = 30×10^4) has been synthesized by the method of free radical polymerization

[22]. Addition of 20% VCL units in the PVP chain is shown to decrease $T_{ph.s}$ of its solution from 170 °C (the calculated value [3]) down to 70 °C. The same effect is observed for copolymers of VP and 2-NN-dimethylamino ethylmethacrylate [23].

Effect of temperature on the structural characteristics of VP-VCL copolymer molecules has been investigated in very dilute aqueous solutions [24]. Relaxation times $\tau_{1\ mm}$, characterizing the intramolecular mobility of macromolecules, were determined with the use of polarized luminescence at stationary light exitation. An increase in $\tau_{1\ mm}$ value indicates an increase in intramolecular hindrance, i.e., macromolecule compactization. It was found that the greater the proportion of VCL units in the copolymer, the greater is an increase in $\tau_{1\ mm}$. The highest increase in $\tau_{1\ mm}$ was observed for the copolymer, containing 85 mol. % VCL units at 43-48 °C. This phenomenon was explained by disturbing syndiotactic sequences of VCL units in the presence of some VP units in the macromolecule. It is interesting that the copolymers were shown to exhibit the higher complexing ability with phenol than PVP and PVCL homopolymers with the maximum for the above copolymer composition.

Many of thermosensitive water-soluble polymers have LCST near 30 °C. Introduction of VP units in the chains of such polymers increases their LCST. For example, this is true for VP-VCL copolymers described above. The behavior of water solutions of VP-NIPAM copolymers, containing 5 and 10% VP units, was also studied [25]. $T_{ph.s}$ of such copolymers was found to be somewhat higher than that of PNIPAM. Formation of monomolecular globules of these copolymers stable in very weak solutions was shown to take place under similar conditions [26]. VP-NIPAM copolymers turned out to be pH-responsive, too. An increase in LCST of the copolymer as compared with that of PNIPAM was found to be higher at pH = 4 than at pH = 7.4 because of the basic character of VP [27].

The $T_{ph.s}$ dependence of random VP and N,N-dimethylacrylamide copolymer solutions on the amide units content is reported [28]. It is shown that $T_{ph.s}$ of the copolymers in 1% water solutions, containing Na_2SO_4 and Na_2CO_3 (0.55M), increase linearly from 7 to 45-49 °C with increasing the PVP content from 0 to100%.,

Recently, PVP derivatives, which have n-C_2H_6 and n-C_4H_9 groups in lactam rings, have been synthesized [29]. These polymers turned out to also exhibit reversible temperature-dependent water solubility. The polymers were applied as stabilizers in formation of "smart" thermoresponsible Au nanoparticle catalysts.

STIMULI-RESPONSIVE HYDROGELS ON THE BASE OF PVP COPOLYMERS

PVP-g-PNIPAM hydrogel was prepared by NIPAM grafting from crosslinked PVP derivative hydrogel beads. The polymerization was initiated by PVP-Br, which was prepared through bromination of pendant allylic groups of the PVP derivative with N-bromosuccinimide. PVP-g-PNIPAM hydrogels are reported to show more rapid temperature-responsive properties as compared with those of hydrogels based on conventional VP-NIPAM copolymers owing to the rapid dehydration of the freely mobile graft chains [30].

Crosslinked pH-sensitive copolymers of PVP and polyacrylic acid (PAA) were prepared by free radical copolymerization of VP and AA in the presence of ethylenglycole dimethacrylate [31].

Preparation of pH-sensitive semi-interpenetrating network (semi-IPA) hydrogel based on hydrogen bond between chemically crosslinked PVP and linear PAA has been reported [32]. Crosslinked PVP beads were obtained by suspension polymerization of VP with AIBN in the presence of NN'-methylene-bis-acrylamide. Physical complexation occurred between PAA and porous PVP beads in the water media. The semi-IPA hydrogel was shown to have excellent pH-sensitivity in the range of pH from 2.2 to 4.0. With increasing acidity, COOH groups of PAA partly dissociate that results in the chain repulsion and formation of hydrogels of higher swelling degree. It is noted that addition of NaCl does not affect much the pH-sensitivity of the hydrogel.

REFERENCES

[1] Dimitrov, I.; Trzebika, B.; Müller, A. H. E.; Dworac, A; Tsvetanov, C. B. *Prog. Polym. Sci.* 2007, vol. 32, 1205-1237.

[2] Aseyev, V. O.; Tenhu ,H.; Winnik, F. M. *Adv. Polym. Sci.* 2006, vol. 196, 1-85.

[3] Kirsh, Yu. E. *Water-soluble Poly-(N-vinylamides): Synthesis and Physico-Chemical Properties*; John Wiley and Sons: Chichester, 1998.

[4] Kirsh, Yu. E. *Prog. Polym. Sci.* 1993, vol. 18, 519-542.

[5] Kumar, A.; Srivastava, A.; Galaev, I. Yu.; Mattiasson B. *Prog. Polym. Sci.* 2007, vol. 32, 1205-1237.

[6] Makhaeva, E. E.; Tenhu, H.; Khokhlov, A. R. *Macromolecules* 2002, vol. 35, 1870-1876.
[7] Sun, T.; King, H. E. *Phys. Rev. E* 1996, vol. 54, 2696-2703.
[8] Sivudu, K. S.; Shailaja D. *Mater. Lett.* 2007, vol. 61, 2167-2169.
[9] Chen, C.; Wang, L.; Jiang, G.; Yu, H. *Res. Adv. Mater. Sci.* 2006, 11, 1-18.
[10] Si, R.; Zhang, Y.-W.; You L.-P.; Yan, C. H. *J. Chem. Phys. Ser. B* 2006, vol. 110, 5994-6000.
[11] Kirsh, Yu. E.; Yanul, N. A.; Kalninsh, K. K.; Maslov, V. G. *J. Molec. Liq.* 1999, vol. 82, 117-130.
[12] Nakhmanovich, B. I.; Pakuro, N. I.; Akhmet'eva, E. I.; Litvinenko, G. I.; Arest-Yakubovich, A. A. *Vysokomolek. Soedin. Ser. B* 2007, vol. 49, 941-944; *Polym. Sci. B* 2007, vol. 49, 136-138.
[13] Güner, A. *J. Appl. Polym. Sci.* 1996, vol. 62, 785-788.
[14] Güner, A.; Ataman, M. *Colloid. Polym. Sci.* 1994, vol. 272, 175
[15] Nakhmanovich, B. I.; Arest-Yakubovich A. A. (to be published).
[16] Pakuro, N. I.; Nakhmanovich, B. I.; Pergushov, D. V. ; Chibirova, F. Kh. *Vysokomolek. Soedin. Ser. A* 2011, vol. 53, 9-14; *Polym. Sci. A* 2011, vol. 53, 6-11.
[17] Colton, R.; Canterford, J. H. *Halides of the First Row Transition Metals*; Wiley and Sons: London, 1969.
[18] Molyneux, P. *Water-Soluble Synthetic Polymers: Properties and Behavior*; CRC Press: Boca Raton. Fl.: 1984, Vol. 2.
[19] Pakuro, N. I.; Yakimansky, A. V.; Chibirova, F. Kh.; Arest-Yakubovich, A. A. *Polymer* 2009, vol. 50, 148-153.
[20] Nakhmanovich, B. I.; Arest-Yakubovich, A. A. (to be published).
[21] Costa, R. O. R.; Freitas, R. F. S. *Polymer* 2002, vol. 43, 5879-5885.
[22] Popkov, Yu. M.; Nakhmanovich; B. I.; Chibirova F. Kh.; Bune, E. V.; Arest-Yakubovich, A. A. *Polym Sci. B* 2007, vol. 49, 155-158.
[23] Nakhmanovich B. I.; Arest-Yakubovich, A. A. (unpublished results).
[24] Anufrieva, E. V.; Gromova, R. A.; Kirsh, Yu; E., Yanul, N. A.; Krakoviak, M. G.; Lushchik, V. B.; Pautov, V. D.; Sheveleva, T. V. *Eur. Polym. J.* 2001, vol. 37, 323-328.
[25] Siu, M. H.; He, C; Wu, C. *Macromolecules* 2003, vol. 36, 6588-6592.
[26] Siu, M. H.; Zhang, G. Z.; Wu, C. *Macromolecules* 2002, vol. 35, 2723-2727.
[27] Dincer, S.; Rzaev, Z. M. O.; Piskin, E. *J. Polym. Res.* 2006, vol. 13, 121-131.

[28] Alencar de Queiroz, A. A.; Gallardo, A.; San Román, J. *Biomaterials* 2000, vol. 21, 1631-1643.
[29] Yan, N.; Zhang, J.; Yuan, Y.; Chen, G.-T.; Dayson, P. J.; Li, Z.-Ch.; Kou, Y. *Chem. Commun.* 2010, vol. 46, 1631-1633.
[30] Jin, S.; Liu, M.; Chen, S.; Gao, C. *Eur. Polym. J.* 2008, vol. 44, 2162-2170.
[31] Devine, D. M.; Higginbottom, C. L. *Eur. Polym. J.* 2005, vol. 41, 1272-1279.
[32] Jin, S; Liu, M.; Zhang, F.; Chen, S.; Niu, A. *Polymer* 2006, vol. 47, 1526-1532.

INDEX

A

absorption spectra, 4, 41
acid, 65, 66, 71, 75, 76, 106, 109
activation energy, 55
adhesives, 65
alcohols, 106
amorphous phases, 80
amorphous polymers, 83
anisotropy, 12, 16, 80
annealing, 23, 24, 41
aqueous solutions, 108
aromatic compounds, 106
aromatic rings, 76, 94
atmosphere, 69, 70
atmospheric pressure, 69
atoms, 3, 4, 13, 14, 25, 91

B

bandwidth, 9, 17, 36
basic research, 57
bending, 70
benzene, 35
biocompatibility, 104
biodegradability, 64, 94, 99
biodegradation, 64, 94
biomass, viii, 63, 66
biotechnology, 104
bonding, 2, 12, 13, 35, 106
bonds, 2, 3, 7, 13, 15, 25, 28, 31, 35, 39, 40, 48, 52
breakdown, 7, 14, 25
breathing, 12
bromination, 109

C

calibration, 7, 69
candidates, 64, 94
carbon, 2, 3, 4, 5, 12, 13, 14, 15, 19, 35, 36, 52, 91
carboxylic acid, 71
catalyst, 68
catalytic effect, 71
cation, 105
chemical, 4, 15, 27, 35, 53, 64, 67, 75, 83, 92
clarity, 36
clusters, 27
collaboration, 57
combined effect, 31
commercial, 67
composition, 50, 57, 92, 94, 98, 99, 108
composting, 94
compounds, viii, 15, 35, 103, 106
compressibility, 16, 25, 47
compression, 4
conduction, 42

configuration, 67, 68, 79, 80, 83, 85, 87, 88, 89, 90, 91, 92, 97
conjugation, 46, 56
contradiction, 56
cooling, 7, 49, 52, 70, 72, 73, 78, 79, 81, 82, 84, 86, 88, 94, 95, 96, 97
correlation, 17, 64, 67, 84, 86, 98
covalent bond, 3, 7, 25, 27, 35, 39, 56
cycles, 43, 44, 45, 46

environment, vii, 4, 63
ester, 67, 75, 91
ethanol, 7
ethylene, 76
evidence, 39, 79, 85
excitation, 6, 27, 32, 33, 41
exciton, 44, 46, 47
experimental condition, 34, 71, 79
exposure, 33

D

decomposition, vii, 2, 6, 49, 50, 51, 52, 53, 54, 55, 57, 107
defects, 17, 41
deformation, 2, 42, 47
degenerate, 17, 36
degradation, 33
degree of crystallinity, viii, 64, 74
dehydration, 109
depth, 2
derivatives, 108
destruction, 3, 15, 25, 40, 48
diamines, 81
diamonds, 51
differential scanning, 48
differential scanning calorimetry, 48
diffraction, 16, 31, 35, 39
dimensionality, 2, 5
dimethacrylate, 109
disorder, 5, 25, 39, 52
dispersion, 22
dissociation, 49, 107
distortions, 56
distribution, 4, 11, 19, 37, 42, 43, 44, 83, 92
doping, 2
drug release, 104

F

feedstock, 64
flexibility, 75, 76, 83, 97, 98
fluorescence, 7, 10, 41, 44
force, 81
formation, viii, 2, 4, 7, 13, 14, 16, 25, 27, 34, 35, 39, 49, 52, 56, 64, 74, 80, 81, 86, 103, 106, 107, 108, 109
fragility, 55, 57
free radical copolymerization, 109
fullerene, vii, 1, 2, 3, 4, 12, 19, 25, 31, 42, 44, 47, 48, 55, 56, 57
fusion, 70

G

gel, 69
gel permeation chromatography, 69
geometry, 5, 35, 91
glass transition, 70, 81, 83, 90
glass transition temperature, 70, 81
graphite, 14
growth, 3

H

hardness, 2, 3, 14, 56
heat capacity, 70
heating rate, 49, 70
helium, 42
history, 70
homopolymers, viii, 64, 66, 70, 75, 92, 94, 96, 108

E

endothermic, 49
energy, vii, 3, 8, 10, 17, 21, 22, 23, 36, 37, 38, 41, 42, 43, 44, 45, 46, 47, 48, 54, 55, 57, 63

Index

humidity, 64
hydrogels, viii, 103, 109
hydrogen, 104, 107, 109

I

ideal, 64, 94
identification, 4, 49
illumination, 2
improvements, 65
impurities, 41
inhomogeneity, 10, 25
initial state, 50
insertion, 80
intermolecular interactions, 75
intrinsic viscosity, 105
inversion, 9
ions, 106
irradiation, 31, 32, 34, 35, 56
islands, 25
isomerization, 71, 81
isomers, 78

K

kinetics, 49, 52
kinks, viii, 64, 80, 85, 88, 98

L

lasers, 6
lattice parameters, 16
lead, 11, 98
light, 2, 34, 35, 99, 108
linear dependence, 22, 55
liquid chromatography, 69
low temperatures, 42, 57
luminescence, 6, 108

M

macromolecular chains, 64, 74, 85, 88
macromolecules, 74, 105, 108

magnitude, 34, 56
majority, 10, 16, 21, 26, 37
mass, 106
materials, vii, 3, 4, 13, 19, 25, 44, 63, 64, 65, 66, 67, 76, 79, 81, 94, 98, 99
matrix, 98
measurements, 7, 9, 16, 26, 32, 34, 36, 37, 41, 42, 47, 54, 55, 70, 105
mechanical properties, viii, 63, 64, 94
media, vii, 109
medicine, viii, 103, 104
melt, 65, 69, 74, 78, 79, 80, 81, 82, 88, 96
melting, 64, 70, 74, 75, 76, 79, 81, 84, 85, 86, 88, 94
melting temperature, 70, 74, 76, 79, 94
metal salts, 105
meter, 69
methacrylic acid, 104
methanol, 7
microcrystalline, 14
microhardness, 14
microorganisms, 66
microscope, 6
mixing, 35
modulus, 16, 56
molar ratios, 69
mold, 70
molecules, 2, 25, 34, 35, 39, 40, 106, 108
monomers, 23, 24, 49, 50, 51, 57, 65, 66, 67, 70, 107

N

nanotechnology, 104
next generation, 64
nitrogen, 5, 69, 70, 78
novel materials, 67

O

oil, 69
oligomers, 2, 17, 18, 49, 50, 51, 52, 53, 54, 57
organic compounds, 106

organize, 96

P

patents, 65
permit, 35, 39
petroleum, 65
pharmacology, 104
phase transformation, 4, 12, 21, 22
phase transitions, vii, 1
phenol, 108
phosphates, 104
phosphorescence, 44
photoluminescence, 6
photons, 2
plastics, 64
pollution, 64
polyamides, 66, 80, 81
polyesters, vii, 63, 64, 65, 66, 67, 75, 76, 78, 80, 81, 83, 98, 99
polystyrene, 69
polyurethanes, 66
preparation, 31, 57, 65, 67
pressure gradient, 19
probability, 83, 86, 87, 89, 90, 92
probe, 4, 7
propylene, 67, 76
protons, 94
prototype, 4
purity, 43, 47, 67, 70

Q

quartz, 7

R

radiation, 2, 35
radical polymerization, 107
reactions, 35, 71
reactivity, 92
recall, 83
recrystallization, 84, 86
redistribution, 18

relaxation, 13
repulsion, 109
residues, 66
resistance, 64, 99
resolution, 6, 26, 35, 39
resources, vii, 63, 64, 65
response, 7, 19, 23, 25, 36, 40, 44
rings, 2, 7, 65, 67, 68, 70, 76, 80, 81, 83, 84, 85, 86, 87, 88, 89, 90, 91, 92, 98, 106, 108
risk, 94
room temperature, vii, 1, 9, 18, 34, 36, 37, 43, 44, 45, 46, 49, 50, 52, 53, 55, 69
rotations, 75
rubber, 70

S

safety, 64
salts, viii, 103, 104, 105
scattering, 4, 5
sedimentation, 107
semiconductor, 41, 42, 47
sensitivity, vii, 109
shear, 4
showing, 9, 13, 19
signals, 74
single crystals, 42
sol-gel, 104
solidification, 9, 19, 36
solubility, 108
solution, viii, 41, 103, 104, 105, 106, 107, 108
solvents, 2
spectroscopy, 4, 92
stability, 4, 9, 16, 32, 36, 40, 48, 54, 64, 76
stabilizers, 108
standard error, 51, 54
stars, 46, 48
state, 2, 15, 19, 22, 24, 31, 34, 35, 39, 41, 44, 49, 50, 52, 54, 55, 56, 57, 78, 79, 80, 81, 83, 98
stretching, 7, 14, 15, 24

structure, vii, 2, 3, 4, 5, 13, 14, 16, 17, 19, 25, 26, 27, 35, 36, 39, 41, 42, 43, 45, 47, 56, 68, 80, 83, 91, 92, 96, 99
substitution, 9, 17, 64, 75, 76
subtraction, 10
swelling, 109
symmetry, 2, 5, 7, 8, 9, 17, 18, 21, 26, 27, 34, 35, 41, 42, 52, 80, 83, 92, 98
syndiotactic sequences, 108
synthesis, 5, 9, 65, 68, 70, 71, 92

T

technology, 2
temperature, vii, viii, 2, 3, 4, 5, 7, 9, 14, 23, 41, 42, 48, 50, 51, 52, 53, 54, 55, 57, 69, 70, 81, 84, 88, 103, 104, 108, 109
tensile strength, 65
terpenes, 65
tetragonal lattice, 13
thermodynamic parameters, 4
thermograms, 75, 79, 80, 82
thermogravimetric analysis, 70
thermoplastics, 64
time use, 52
titanium, 67
total energy, 54, 55
toxicity, 64, 104
trade, 94

transformation, vii, 1, 15, 19, 23, 24, 26, 30, 32, 34, 35, 36, 38, 39, 41, 45, 49, 56
transition metal, viii, 103, 105
transition temperature, 94
treatment, vii, 2, 3, 4, 5, 30, 41, 49, 50, 51, 52, 53, 54, 55, 57

U

uniform, 3, 7, 23, 25, 36

V

vacuum, 70
variations, 79
vibration, 7, 12, 24
viscosity, 69

W

waste, 64
water, vii, viii, 69, 70, 99, 103, 104, 105, 106, 107, 108, 109

Y

yield, 41, 44